AF678516

Structural Health Monitoring Using Emerging Signal Processing Approaches with Artificial Intelligence Algorithms

Structural health monitoring is a powerful tool across civil, mechanical, automotive, and aerospace engineering, allowing the assessment and measurement of physical parameters in real time. Processing changes in the vibration signals of a dynamic system can detect, locate, and quantify any damage existing in the system. This book presents a comprehensive state-of-the-art review of the applications in time, frequency, and time-frequency domains of signal-processing techniques for damage perception, localization, and quantification in various structural systems.

Experimental investigations are illustrated, including the development of a set of damage indices based on the signal features extracted through various signal-processing techniques to evaluate sensitivity in damage identification. Chapters summarize the application of the Hilbert–Huang transform based on three decomposition methods such as empirical mode decomposition, ensemble empirical mode decomposition, and complete ensemble empirical mode decomposition with adaptive noise. Also, the chapters assess the performance and sensitivity of different approaches, including multiple signal classification and empirical wavelet transform techniques in damage detection and quantification. Artificial neural networks for automated damage identification are introduced.

This book suits students, engineers, and researchers who are investigating structural health monitoring, signal processing, and damage identification of structures.

Chunwei Zhang is a Chair Distinguished Professor at Shenyang University of Technology. He is the Founding Director of the Multidisciplinary Center for Infrastructure Engineering (MCIE) at Shenyang University of Technology, and the Founding Director of the Structural Vibration Control (SVC) Group at Qingdao University of Technology, China. His research achievement and worldwide impact have been highly recognized by the international academia society, as evidenced by the continuous inclusions into the prestigious rankings, such as the Clarivate Highly Cited Researcher, Elsevier Most Cited Chinese Researcher, and Stanford World Top 2% Scientists, among others. Apart from publications, his inventions have been implemented in the engineering practice as evidenced by the active control system for the Canton Tower structure. He is also a commended author of CRC Press published books and proceedings.

Asma A. Mousavi is a PhD graduate in Civil and Structural Engineering from Qingdao University of Technology.

Structural Health Monitoring Using Emerging Signal Processing Approaches with Artificial Intelligence Algorithms

Chunwei Zhang and
Asma A. Mousavi

CRC Press
Taylor & Francis Group
Boca Raton London New York

CRC Press is an imprint of the
Taylor & Francis Group, an **informa** business

MATLAB® and Simulink® are trademarks of The MathWorks, Inc. and are used with permission. The MathWorks does not warrant the accuracy of the text or exercises in this book. This book's use or discussion of MATLAB® or Simulink® software or related products does not constitute endorsement or sponsorship by The MathWorks of a particular pedagogical approach or particular use of the MATLAB® and Simulink® software.

First edition published 2025
by CRC Press
4 Park Square, Milton Park, Abingdon, Oxon, OX14 4RN

and by CRC Press
2385 NW Executive Center Drive, Suite 320, Boca Raton FL 33431

© 2025 Chunwei Zhang and Asma A. Mousavi

CRC Press is an imprint of Informa UK Limited

The right of Chunwei Zhang and Asma A. Mousavi to be identified as authors of this work has been asserted in accordance with sections 77 and 78 of the Copyright, Designs and Patents Act 1988.

All rights reserved. No part of this book may be reprinted or reproduced or utilised in any form or by any electronic, mechanical, or other means, now known or hereafter invented, including photocopying and recording, or in any information storage or retrieval system, without permission in writing from the publishers.

For permission to photocopy or use material electronically from this work, access www.copyright.com or contact the Copyright Clearance Center, Inc. (CCC), 222 Rosewood Drive, Danvers, MA 01923, 978-750-8400. For works that are not available on CCC please contact mpkbookspermissions@tandf.co.uk

Trademark notice: Product or corporate names may be trademarks or registered trademarks, and are used only for identification and explanation without intent to infringe.

British Library Cataloguing-in-Publication Data
A catalogue record for this book is available from the British Library

ISBN: 978-1-032-80613-6 (hbk)
ISBN: 978-1-032-81282-3 (pbk)
ISBN: 978-1-003-49904-6 (ebk)

DOI: 10.1201/9781003499046

Typeset in Sabon
by codeMantra

Contents

Preface

Damage detection and operational safety assessments of strategic civil infrastructures, e.g. bridges, have always been vital challenges for structural health monitoring (SHM). Nowadays, the time-frequency-based methods of signal processing as a part of vibration characteristic-based techniques are increasingly utilized by researchers. Among the various parts of existing approaches, signal-processing techniques are known as the key components that are frequently used in scientific investigation and engineering practice. Most of the primary signal-processing approaches are based on the fact that the signal is assumed to be linear and stationary. However, real-world signals are often inherently nonstationary and nonlinear in nature. Since the time–frequency techniques can supply both time and frequency details of a signal, they are appropriate for processing high-order nonlinear and nonstationary signals. Therefore, selecting the optimal and efficient signal-processing techniques becomes one of the important tasks for SHM. This book aims to evaluate the performance of the most recently introduced or developed signal-processing techniques combined with machine learning algorithms such as artificial neural networks in comparison with the performances of traditional approaches for damage detection, localization, and classification of a Benchmark laboratory steel truss bridge structural model.

This book appears to occupy a unique and specialized niche within the realm of published books. Its focus on evaluating signal processing approaches, combined with machine learning algorithms, for damage detection, localization, and classification of a laboratory steel truss bridge structural model sets it apart from more generalized literature. It caters specifically to a selective audience of students, engineers, and researchers in the science and engineering fields, particularly those involved in Civil Engineering discipline.

Compared to other published books, which may cover a broader range of topics within Civil Engineering or structural analysis, this book delves deeply into the application of time, frequency, and time-frequency signal processing techniques for damage assessment. This specialized approach allows it to provide comprehensive insights and state-of-the-art reviews tailored to the needs of postgraduate students and researchers.

Foreword

Structural health monitoring (SHM) has become an important and hot topic for decades in various fields of civil, mechanical, automotive, aerospace engineering, etc. Estimating the health condition and understanding the unique characteristics of structures through assessing measured physical parameters in real-time is the major objective of SHM. As a result, signal processing becomes an essential and inseparable approach of vibration-based SHM investigation. The basic goal of using signal processing is to perceive the changes or identify the severity of changes from the vibration signals of the dynamic system to detect, locate, and quantify any damages existing or occurring in the system. Furthermore, vibrations of complex structures such as bridges mostly present nonlinear and nonstationary behaviors. One of the serious issues of traditional signal-processing techniques in analyzing the responses of real-life structures is related to the presentation of fundamental information of nonlinear, nonstationary, and noisy signals with closely spaced frequencies. To overcome this difficulty, numerous studies have been recently carried out to explore proper time-frequency signal-processing techniques to efficiently distinct high-resolution representations for nonlinear characteristics of analyzed signals. Despite the existing extensive reviews on vibration-based signal-processing techniques in time and frequency domains for SHM purposes, there exists no study in categorizing the signal-processing techniques based on the feature extraction in time, frequency, and time-frequency representations simultaneously.

To fill this gap, firstly this book presents a comprehensive state-of-the-art review on the applications of time, frequency, and time-frequency signal-processing techniques for damage detection, localization, and quantification in various structural systems in Chapter 1. In addition, the theoretical background is overviewed for relevant specific signal-processing approaches adopted in the present work to analyze the experimental measurements. Moreover, the significance and purpose of this book are also described in this chapter.

The research contribution of this book is declared in Chapter 2. The case study model and experimental setup are introduced in detail. The research methodology for the implementation of combined signal-processing techniques and machine learning in damage identification, localization, and quantification is described in this chapter. In addition, a series of damage indices based on the signal features extracted by various signal-processing techniques are proposed to evaluate the sensitivity in damage identification. Finally, systematical comparative studies are carried out to validate the aforementioned signal-processing techniques.

In line with the studies presented in Chapter 3, the application of HHT based on three decomposition methods including empirical mode decomposition (EMD), ensemble empirical mode decomposition (EEMD), and complete ensemble empirical mode decomposition with adaptive noise (CEEMDAN) are investigated and compared in terms of damage detection and quantification. Furthermore, the performance and the sensitivity of multiple signal classification (MUSIC) and the combined CEEMDAN-MUSIC techniques for damage detection and quantification are investigated in this chapter. The above methods are further compared with traditional methods in frequency domains such as FFT, PSD, and FDD.

In Chapter 4, the application of EMD, EEMD, and CEEMDAN for damage localization is assessed and compared. In addition, MUSIC, CEEMDAN-MUSIC, FFT, PSD, and FDD techniques are evaluated and compared for damage localization of a truss bridge structure. The contribution and main finding of this part are explained at the end of this chapter.

In Chapter 5, the efficiency of artificial neural network (ANN) is evaluated to automate the damage identification approach using CEEMDAN Hilbert transform and empirical wavelet transform (EWT) techniques with the purpose of effectively reducing the human intervention and speeding up the process of structure diagnosis. Detection, quantification, and localization of the damage of the truss bridge model when subjected to white noise and impact loads are assessed by the EWT-ANN method.

Finally, Chapter 6 presents strategic conclusions and main findings of this book as well as discusses some recommendations for future work.

Keywords: Structural health monitoring; Signal processing techniques; Feature extraction; Artificial Neural Network; Damage detection.

Acknowledgments

The authors would like to acknowledge the following financial support by the National Natural Science Foundation of China (Grant No. 52361135807), the Ministry of Science and Technology of China (Grant No. 2019YFE0112400), the National Foreigner Expert Bureau Supporting Scheme, the Department of Science and Technology of Shandong Province (Grant No. 2021CXGC011204), the Liaoning provincial key laboratory of Safety and Protection for Infrastructure Engineering, and the Central Government and Liaoning Provincial Discovery Project.

Chapter 1

Introduction

1.1 BACKGROUND, SIGNIFICANCE AND PURPOSE

1.1.1 Background

Strategic and complex civil structures and infrastructures are one of the signs of a modern and developed society, which is utilized in transportation, economics, and the protection of people and products. Due to the importance of the aforementioned utility, deterioration, and degradation of civil structures are some of the highlight concerns all over the world. Besides, they are exposed to severe hazardous occurrences and continuous ambient vibrations, and natural phenomena such as earthquakes, floods, tornadoes, corrosion, traffic loads, material aging, impact, environmental corrosions, etc. Therefore, monitoring and evaluating the performance of structures is an essential demand to enhance structural integrity and safety. Structural health monitoring (SHM) and research on structural damage detection approaches have become an interesting and fast-growing topic for decades in various scientific fields of aerospace, civil, and mechanical engineering. Various types of sensor devices such as acoustic, pressure, temperature, force, and accelerometer sensors are used in SHM to measure the signal data.

Estimating the health condition and the structural characteristics of the structure through assessing measured physical parameters in real time is the goal of SHM. In recent decades, the researchers in SHM have been investigating the development of cost-effective and vigorous monitoring approaches through utilizing and integrating novel technologies from different information and engineering technology methods. Furthermore, various data mining, visual tools, and statistical computing are feasible to employ and integrate into the monitoring and processing solution. Especially, in recent years, SHM approaches focused on vibration-based monitoring due to the ability to detect the invisible damages of the internal section of the structures through the variation of structural characteristics such as stiffness that can be manifested in the vibration measurements. Therefore, utilizing robust signal processing techniques to indicate the

DOI: 10.1201/9781003499046-1

changes of vibration response of the structures is of paramount importance in damage localization, quantification, and detection process [1]. To this end, extracting features from the sensed data before and after damages is the objective of signal processing techniques. The aim of employing the signal processing on the response signal from the structures is to extract the features to define whether it is damaged or not, the type and severity of the damage, and its location [2]. The signal processing techniques should deal with the complexity of structural response under the various dynamic loads, noisy environments, and complicated structure models challenging the feature extraction process in SHM of different systems [3].

Compared to the feature extraction performance of traditional signal processing techniques that necessitate updating the measured data, modern data-based techniques mainly employ statistical pattern recognition approaches. These data-based techniques in SHM consist of feature extraction based on the inherent characteristics of the studied systems, and decision-making based on the analyzed data that can be significantly affected by substantial variability in environmental conditions. The effects of environmental and operational variability (EOV) conditions such as ambient temperature, wind speed, humidity, traffic loading, operational speed, mass loading, etc., on the signals collected from the structure of interest are very important and still a major challenge in the SHM community. In performing SHM of structures in the real world, many uncertainties such as environmental variability, noises, and ambient excitations may disturb the damage identification procedure. These conditions can directly affect the performance of vibration-based damage diagnosis techniques at all levels such as early damage detection, damage localization, and damage quantification. To overcome these negative effects of EOV conditions, several research studies have been conducted previously. Sarmadi et al. [4] proposed an ensemble learning-based technique using a non-generative sequential algorithm for SHM of a wooden bridge under different environmental conditions. In this method, the negative environmental effects on the performance were significantly mitigated by adopting a nonparametric data-based framework. The superiority of the proposed approach was concluded in detecting the damage compared to the traditional vibration-based damage detection techniques. Entezami et al. [5] introduced a data-driven technique for SHM of a cable-stayed bridge based on the analysis of time-series (TS) statistical and high-dimensional features. In the proposed method, a multifunctional calculation method from the combined actions of a series of calculation procedures including an automated identification algorithm, an improved order determination technique, and a hybrid distance identifier approach was implemented for detecting the damage. In line with this study, a new data-driven approach based on the implementation of a series of computational procedures including segmentation, pairwise distance calculation, an iterative algorithm by using a multidimensional scaling procedure, a matrix vectorization process, and a Euclidean norm calculation method was proposed by Entezami et al. [6] for damage detection of an IASC-ASCE benchmark structure under ambient excitations. The superiority of the proposed method that was based on the extraction of high-dimensional features including TS statistical features was concluded in detecting the damage compared to the results from the extraction of traditional multidimensional features having lower sensitivity to the existence of the damage.

Most of signal processing algorithms that have been frequently used in vibration-based SHM with specialization in damage detection approach are reviewed in this research study roughly in chronological order. This study particularly presents a state-of-the-art review study of the vibration-based signal processing and feature extraction methods employed for SHM purposes. The vibration signals of civil structures such as bridges, buildings, and frames with various materials assessed through the signal processing techniques are presented. In this study, two categories of signal processing methods related to the domains of extracted features are reviewed namely (i) time-domain and (ii) frequency domain. Frequency-based signal processing techniques can be categorized into parametric and nonparametric methods. The procedure of parametric signal processing techniques is based on the extraction of modal parameters such as natural frequencies, mode shape, and damping ratio of the analyzed signals [7]. These techniques are commonly applied on signals with known sources. However, nonparametric techniques are employed when there is no prior knowledge about signal source such as signal responses measured under moving traffic loads, speed-rated signals, or occupancy signals [8–12]. The frequency-domain signal processing techniques can be also classified into four types in terms of their damage detection performance based on the changes of modal parameters including damage detection based on the changes of natural frequencies, mode shapes, modal damping ratios, and updating the structural model parameter. Among these techniques, frequency-based methods are more commonly utilized for structural damage detection purposes. However, frequency-based techniques have disadvantages in damage localization and quantification [13,14]. The time-domain signal processing techniques mainly refer to statistical TS methods for SHM. The statistical TS techniques perform based on the random extraction procedure of vibration signals, developing the statistical model of the analyzed signals, and the statistical estimation of the state (healthy or damaged) of the analyzed system [15–20]. The statistical TS techniques that are data-based techniques have more advantages over general vibration-based techniques that is the signal faults can be accurately quantified by statistical time series without any need for visual inspection of the physical model of the studied system. Despite multiple review studies published on vibration-based structural damage detection, there exists no study in categorizing the signal processing techniques based on the features extraction procedure with specialization in damage detection approach that belongs to time and frequency domains.

1.1.2 Significance of the book

Damage detection and operational safety assessments of bridges as the strategic civil infrastructures have always been vital challenges in SHM. Nowadays, the time-frequency-based methods of signal processing as a part of vibration characteristic-based techniques are increasingly utilized by researchers. Among the various parts of existing approaches, signal processing techniques are known as the key components that are frequently used in scientific and engineering fields. Most of the primary signal processing approaches are based on the fact that the signal is linear and stationary. However, real-world signals are often nonstationary and nonlinear. Since the time-frequency techniques can supply both time and frequency details of a signal, they are appropriate to process nonlinear and nonstationary signals. However, selecting the optimal and efficient signal processing techniques is one of the most important topics in SHM. Therefore, the main objective of this book is to evaluate the performance of novel and the most recent signal processing techniques such as EMD, EEMD, CEEMDAN, Hilbert Huang transform (HHT), EWT, and MUSIC in combination with machine learning algorithms such as artificial neural networks (ANNs) in comparison with the performances of traditional approaches in damage detection, localization, and classification of a laboratory steel truss bridge model. To do this, a series of damage indices based on the signal energy feature and a series of statistical time-dependent features including root mean square (RMS), shape factor, kurtosis, and entropy have been rarely used in the damage detection process of complex civil structures, are proposed. In addition, adopting ANN as a supervised machine learning with the purpose of automatic signal analysis and damage identification can effectively speed up the process of structure diagnosis and reduce the human factor. The main advantage of ANNs compared to other machine learning algorithms is that they allow the nonlinear and nonparametric modeling of complex systems with large datasets simply and adaptively without requiring explicit mathematical representations or without requiring exhaustive experiments. However, most other statistical methods are parametric models that need a high background in statistics. Hence, the use of ANN and its hybridization with signal processing techniques are more popular for SHM applications compared to other intelligent algorithms.

To this end, first, a comprehensive state-of-the-art review on the applications of signal processing techniques in time, frequency, and time-frequency domains for damage detection, localization, and quantification in various structural systems. The progressive trend of time-frequency analysis methods is reviewed by summarizing their advantages and disadvantages when utilized for different applications in various structural and mechanical systems. Then, the capability of several damage indices based on signal features including energy, instantaneous amplitude (IA), unwrapped phase, and instantaneous frequency (IF) extracted through applying the Hilbert transform on decomposed IMFs by selected signal processing techniques was explored. These features have not been evaluated all at once in damage detection procedures in civil engineering applications.

In addition, a series of statistical time-domain features including RMS, shape factor, kurtosis, and entropy was extracted to damage quantification and localization of a scaled truss bridge model subjected to white noise and impact loading excitations which has not been reported in the literature. In fact, from an extensive review on the evaluation of these features, it was concluded that the change of the structure stiffness (i.e., the occurrence of damage) extremely affects the output results based on the aforementioned features. Therefore, these signal features are utilized in damage evaluation and identification through the proposed methodology.

1.1.3 Purpose of research

The main objective of this study is to evaluate the performances of multifarious structural damage identification approaches based on the combination of novel signal processing techniques such as CEEMDAN, MUSIC, and EWT with ANN technique, which can guarantee the robustness, reliability, and efficiency of the detection process. The specific objectives of this book are:

i. To investigate the advantages of an automated damage detection approach from the hybrid application of CEEMDAN and an improved EMD-based technique and ANN named CEEMAN-HT-ANN for SHM of a steel truss bridge which has not been reported in the literature.
ii. To compare the performances of three generations of HHT-based techniques (i.e., EMD, EEMD, and CEEMDAN) based on extracted features for the first time in detecting the damage of the truss bridge, comparatively.
iii. To explore the validity of a proposed hybrid damage detection approach based on the combination of EWT and ANN techniques named EWT-ANN in the damage quantification and localization of a scaled truss bridge model subjected to white noise and impact loading excitations.
iv. To experimentally assess the performance of a hybrid damage detection approach from the combined use of MUSIC algorithm with CEEMDAN named CEEMDAN-MUSIC approach compared to the performance of pure MUSIC and several frequency-domain techniques such as Fast Fourier Transform (FFT), power spectral density (PSD), and Frequency Domain Decomposition (FDD) in identifying the location and quantifying the damage severity of a scaled truss bridge model.
v. To evaluate the presence, location, and severity of damage using four extracted signal features including energy, IA, unwrapped phase, and instantaneous frequency (*IF*) extracted from EMD-based techniques in both time and frequency domains all at once.

vi. To perform a comparative study on the effectiveness of a series of statistical time-domain signal features including RMS, shape factor, kurtosis, and entropy for damage quantification and localization of a scaled truss bridge model which has not been reported in the literature.
vii. To evaluate the accuracy of a series of improved damage indices based on the combinations of two statistical time-history features including kurtosis and entropy features with the energy and IA features of the analyzed signal that has not been reported in SHM.

1.2 TIME-DOMAIN SIGNAL PROCESSING TECHNIQUES

The statistical TS models which are divided into linear and nonlinear statistical models [20] are commonly used to extend an approximate mathematical model based on the input-output measurements. The linear statistical TS models are employed in SHM and condition assessment approaches of structures when subjected to various dynamic loadings. The experimental and numerical analysis demonstrated the efficiency of linear TS models in damage diagnosis methodologies. The linear statistical TS models can be categorized into stationary methods such as autoregressive model (AR), moving average model (MA), autoregressive moving average (ARMA) model; and nonstationary methods such as vector autoregressive (ARV) model, the autoregressive model with exogenous input (ARX), autoregressive integrated moving average (ARIMA), and autoregressive moving average with exogenous inputs (ARMAX). Time-domain signal processing techniques have been utilized across various fields, such as biomedical engineering, mechanical engineering, robotics, etc. However, the application of signal processing methods in SHM with specialization in damage detection of civil engineering structures and infrastructures. Table 1.1 presents a list of outstanding research studies in the application of linear statistical TS models in SHM. Also, the time-domain techniques and the type of analyzed structures including numerical, laboratory, and real-life structures in the selected research studies are categorized.

1.2.1 Autoregressive model

The TS models are commonly used for damage detection of structures under environmental and operational conditions based on the feature extraction approaches associated with operating each model for experimental TS data. In the AR model, the TS-based feature extraction procedure is operated on the measured experimental data for the estimation of future data values using their past values and capturing the residual errors. The AR model performs the damage detection procedure using the estimated parameters that are sensitive to the damage characteristics while being insensitive to the

Table 1.1 Summary of using time-domain signal processing techniques in damage detection of civil structures

Study	*Technique*	*Analyzed structures*	*Level of damage detection*
Sohn et al. [21]	AR	A reinforced-concrete column (L)	Existence
Figueiredo et al. [22]	AR	A three-story base-excited frame (L)	Existence
Vamvoudakis-Stefanou et al. [23]	AR	A composite beam (L)	Existence and location
Jayawardhana et al. [24]	AR	A steel beam (R)	Existence, location, and severity
Lautour et al. [25]	AR	ASCE benchmark	Damage classification
Gul and Catbas [26]	AR	A steel beam and steel grid structure. (L)	Existence, location, and severity
Pamwani et al. [27]	AR	A streel frame (L)	Existence, and severity
Mei et al. [30]	AR	A five-story shear-building (N and L)	Existence, location, and severity
Carden and Brownjohn [31]	ARMA	A four-story frame (L)	Existence
Zheng and Mita [32]	ARMA	A five-story building (N, L)	Existence, location, and severity
Krishnan Nair and Kiremidjian [35]	ARMA	ASCE four-story building (L)	Existence, and severity
Shi et al. [36]	ARMA	A cantilever beam and Bridge model (L)	Existence
Zhang et al. [37]	ARMA	A three-floor shear structure (L)	Existence
Bao et al. [38]	ARMA	Subsea soil-pipeline-fluid (N)	Existence and location
Bernal et al. [46]	ARX	An aluminum beam (N)	Existence
Gul and Catbas [49]	ARX	A steel grid model (L)	Existence and location
Li and Ren [51]	AR-ARX	A building (R)	Existence
Lakshmi and Rao [53]	AR-ARX and ARV	A steel beam (N, L)	Existence, location, and severity
Krishnasamy et al. [54]	AR-ARX	A RCC beam (L)	Existence and location
Poulimenos and Salellariou [55]	ARX	A composite beam (L)	Existence
Tatsis et al. [58]	ARX	A Spring-Mass-Damper System (N)	Existence and location
Li et al. [68]	ARV	A RC beam (L)	Existence and location

(*Continued*)

Table 1.1 (Continued) Summary of using time-domain signal processing techniques in damage detection of civil structures

Study	*Technique*	*Analyzed structures*	*Level of damage detection*
Mattson and Pandit [71]	ARV	An eight-degree of a system (L)	Existence and location
Okasha et al. [73]	ARV	A high-speed naval craft (L)	Existence and location
Bodeux and Golinval [74]	ARMAV	Simply supported beam (N) and Steel-Quake' structure (L)	Existence and location
Omenzetter et al. [40]	ARIMA	Singapore-Malaysia Second Link Bridge (R)	Existence and location
Chen et al. [42]	ARIMA	Hurley Bridge (R)	Existence
Posenato et al. [43]	ARIMA and ARMA	A full-scaled bridge (R)	Existence, location, and severity
Ling and Mahadevan [44]	ARIMA	A Rotorcraft (L)	Existence
Lakshmi et al. [53]	ARMAX	An RCC beam (L)	Existence and location
Do and Gul [61]	ARMAX	A four-story shear (L)	Existence, location, and severity
Lakshmi and Rao [62]	ARMAX	Beam (N), eight-DOF benchmark (L), and FEM 12-DOF steel frame (N)	Existence and location
Mei et al. [63]	ARMAX	A three-dimensional RC frame (L)	Existence
Xie and Mita [64]	ARMAX	Ten-story shear (N) and five-story frame (L)	Existence and severity
Gislason and Gul [65]	ARMAX	Four-story steel and RC building (N)	Existence, location, and severity
Saaed and Nikolakopoulos [66]	ARMAX	Ten-story benchmark steel frame (L)	Existence and severity

Note: N, numerical; L, Laboratory; R, real-life.

variations of operational- and environmental-related factors. There exist many research works employing TS analyzed based on the AR model as a feature extraction method for SHM of civil engineering structures. Sohn et al. [21] utilized the coefficients of autoregressive (AR) models as a statistical process control (SPC) technique to identify the presence of damage. X-bar control charts are used to analyze the sensitive features of different levels of damage from a concrete column model. A comparative study by

Figueiredo et al. [22] identified four different methods including Akaike information criterion, partial autocorrelation function, root means squared error, and singular value decomposition which are used in the AR model to extract the features that are sensitive to the damage from the acceleration response of a three-story frame model. The performance of different statistical pattern recognition methods was investigated for SHM of various laboratory-scaled structures. To this end, the autoregressive TS modeling in the combination with Mahalanobis distance feature as the statistical damage indicator. Stefanou et al. [23] investigated Two Multiple Model (MM) methods based on the autoregressive parameter vector, including Unsupervised Multiple Model AR (U–MM–AR) and its PCA-enhanced version (U–PCA–MM–AR) methods as a vibration-based damage detection methodology on a composite beam laboratory model. A damage-sensitive feature according to the computation of AR model coefficients as a structural damage detection system was introduced by Jayawardhana et al. [24]. The experimental results of a steel beam model structure using wireless sensor networks demonstrated the novelty of the damage identification algorithm. Lautour et al. [25] investigated a damage classification for the use of SHM using AR models to fit the time data of two experimental models including a three-story bookshelf model and the experimental SHM ASCE benchmark structure. The extracted coefficients of AR models were utilized as a damage indicator that was used to feed the ANN to classify damage cases. Gul and Catbas [26] investigated pattern recognition techniques in various laboratory structures for structural damage identification. In this study, the AR model in combination with Mahalanobis distance-based algorithm was used to detect various types of structural damages on two experimental test structures including simply supported steel beam and highly redundant steel grid structure as shown in Figure 1.1.

An autoregressive damage detection methodology was proposed by Gharehbaghi et al. [27] to identify the damage in two building models subjected to ambient and forced vibrations using a series of statistical indices based on a best-uncorrelated features selection (BUFS) algorithm. This algorithm was able to produce a set of predictors for different states of the structure with the lowest correlations. The capability of the proposed method was concluded even in a noisy environment. A time-varying autoregressive (TVAR) model combined with the K-means-clustering technique was proposed by Pamwani et al. [28] to detect the damage in a streel frame structure (see Figure 1.2) when exposed to ambient white noise excitation. The successful performance of the introduced method in quantifying the damage was concluded. Cheng et al. [29] used a hybrid AR/ARCH model to experimentally identify the damage in a three-story structure and an eight-story shear building structure in laboratory scales by proposing a new damage index based on the second-order variance indicator (SOVI).

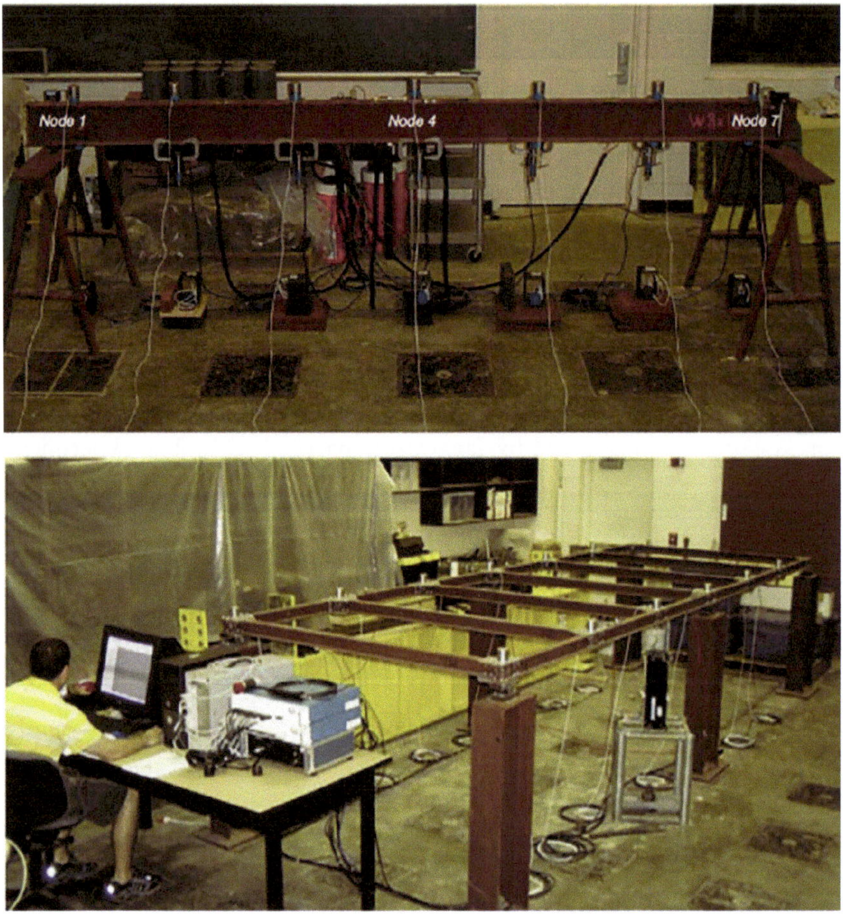

Figure 1.1 The instrumented steel beam and steel grid model test setup [26].

The advantages of the algorithm using SOVI in identifying the nonlinear damage were revealed compared to a traditional linear damage index based on the cepstral metric indicator (CMI). A new output-only damage indicator based on optimal sub-pattern assignment distance and AR model is proposed by Mei et al. [30]. The modal parameter is utilized as a dynamic characteristic to find the change of five-story shear case study. Despite extensive use of AR model in damage identification of the structures based on their modal information such as natural frequencies and damping ratios due to its low computational costs, this technique has a serious deficiency in detecting the location of the damage since the AR lacks spatial information about the structural vibration response.

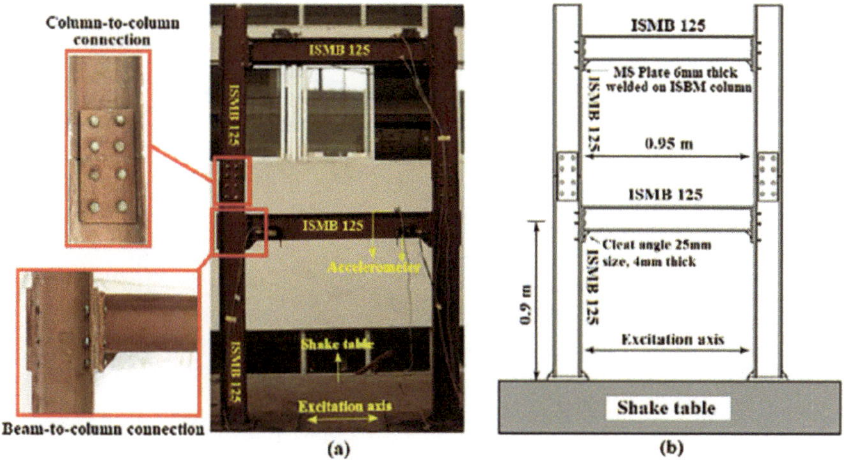

Figure 1.2 (a) Experimental setup and (b) schematic sketch of steel frame [28].

1.2.2 Autoregressive moving average (ARMA)

The ARMA model is another linear TS model which can be employed for dynamic vibration responses of structures to predict the future data based on their past values. This model is integrated with the AR and MA models. Similar to the AR model, the basis of the procedure operated by the ARMA model is to estimate the modal information of the vibration responses such as natural frequencies and damping ratios. However, this model has more advantages in the classification of the time series compared to the AR model. Carden and Brownjohn [31] presented a time-domain statistical classification approach based on the coefficients of ARMA to be fed to the unsupervised classifier to monitor the extraordinary change in the ASCE benchmark frame model. Furthermore, the Z24 bridge and the Malaysia–Singapore Second Link bridge assessed the proposed algorithm as the experimental case studies models. The result demonstrated the capability of the technique to identify structural damages. The performance of the ARMA time series algorithm was evaluated by Nair et al. [32] to localize and detect the existence of damage in the ASCE benchmark structure (see Figure 1.3) utilizing a novel damage-sensitive feature. It was revealed that the introduced methodology has a privilege in locating the damage, in addition, to classifying its severity compared to previous statistical signal processing algorithms.

Zheng and Mita [33,34] utilized autoregressive moving average (ARMA) and AR models in damage localization and quantification of laboratory experimental and numerically simulated data, respectively. The distance

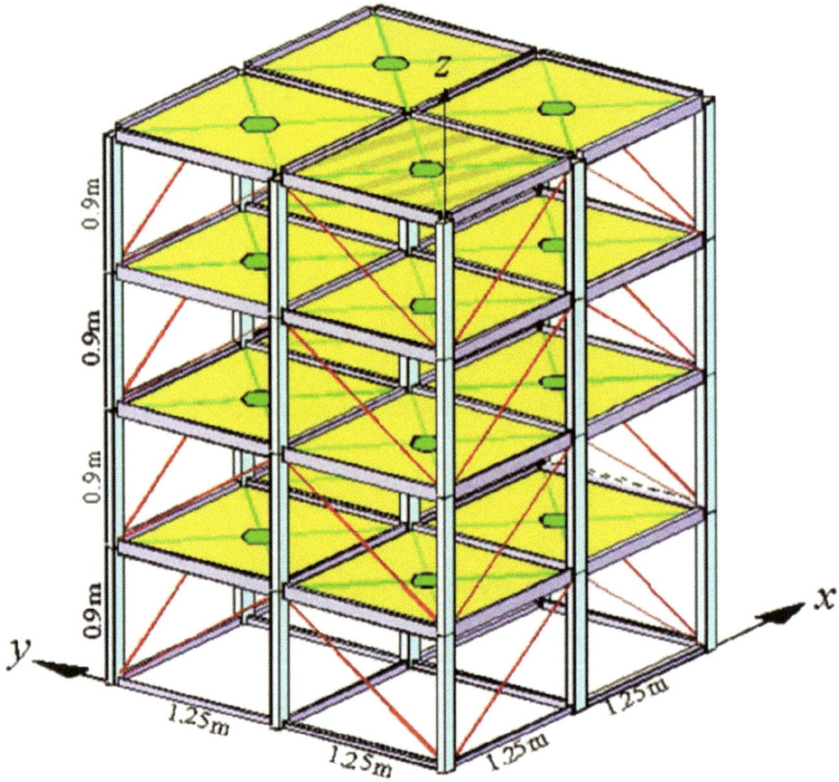

Figure 1.3 Diagram of the ASCE benchmark structure [32].

measures of AR and ARMA models including Itakura distance, cepstral distance, and subspace angle distances are defined as sensitive indicators for damage assessment. Furthermore, a pre-whitening filter was used to remove the correlations in excitations. Nair et al. [35] designed a time series damage identification based on ARMA and Gaussian Mixture Model (GMM) to extract the first three autoregressive coefficients as a feature vector and model the feature vector to calculate the damage indicators, respectively. A numerical model of the ASCE Benchmark four-story braced frame is utilized to verify the efficacy of the introduced algorithm. Shi et al. [36] explored a well-established extension of the cointegration method called the TBATS (Trigonometric, Box-Cox transform, ARMA errors, Trend, and seasonal components) model to monitor the seasonal effects. To this end, a synthetic cantilever beam and National Physical Laboratory (NPL) Bridge model case studies were investigated to evaluate the efficiency of the introduced methodology. The result demonstrates the effect of the TBATS model to extract the daily cycles from the heteroscedastic data.

A non-modal parametric method from the combination of a regularized autoregressive moving average (ARMA) time series model with a virtual impulse response function was proposed by Zhang et al. [37] for damage diagnosis of a three-degrees-of-freedom chain structure and a three-floor shear structure under environmental excitation using. It was found that the regularized ARMA model can precisely obtain the feature of signal and efficiently identify the structural damage.

Bao et al. [38] designed an integrated ARMA model and Partial Auto-correlation Function (PAF) method to extract the damage feature vectors and optimal AR model order, respectively. In this study, the Mahalanobis distance of the different damaged states and the baseline situation is defined as a damage index (DI) to damage localization and detection. The dynamic acceleration vibration of an FE model subsea soil-pipe-line-fluid system simulated by ANSYS software, which was used as a numerical case study, proved the sensitivity of the proposed methodology to detect the damage. Georgantopoulou and Fassois [39] successfully used a parametric statistical damage identification technique based on the combined time-dependent Auto Regressive Moving Average (TARMA) model and PSD to quantitatively damage identification in a laboratory-scale composite beam model.

1.2.3 Autoregressive integrated moving average

The ARIMA model is an improved model of AR that adopts a series of parameters as the damage indices. Due to the drawbacks of the ARMA model in the analysis of noisy data, this model is not able to efficiently obtain the nonlinear characteristics of the analyzed data of structures. In contrast, the ARIMA models present advantages in analyzing the nonstationary and nonlinear data by taking a three-part procedure composed of AR, integrated, and MA parts, respectively. The following research works present outstanding applications of the ARIMA model in the damage detection of civil engineering structures. Omenzetter et al. [40] used the ARIMA model to analyze the recorded strain signals of the major bridge construction. The coefficients of the ARIMA model are identified using an adaptive Kalman filter. By observing the results, changes in the values of the ARIMA coefficients can be revealed when unusual structural changes or damage occur. Xin et al. [41] presented a new methodology based on the combination of Kalman filter, ARIMA, and GARCH to improve the bridge structure behavior prediction. To denoise the raw data, analyze and predict the bridge deformation, and further improve the accuracy of the prediction Kalman filter, ARIMA, and the nonlinear recursive GARCH model are established, respectively. The simulation results from the deformation monitoring system named Global Navigation Satellite System (GNSS) demonstrated the efficiency of the introduced deformation prediction model. The usage of ARIMA in SHM of transportation infrastructure was explored by Chen et al. [42]. A two-stage methodology based on estimation regression and ARIMA-GARCH models is proposed on the strain and displacement data from the Hurley Bridge. The results reveal that the significant components of the measurements including the persistent displacement and trends in the strain data indicate the deterioration of the bridge supports and changes in the material properties, respectively.

A comparative study of two statistical techniques based on (i) moving principal component analysis (MPCA) and (ii) robust regression analysis including ARMA, Box–Jenkins, and seasonal ARIMA are assessed in damage detection of civil structures by Posenato et al. [43]. The results demonstrate the superiority of these methods to detect sudden changes in the behavior of the full-scale bridge. Furthermore, for monitoring noisy environments, MPCA performs better than auto-regressive analysis. Ling and Mahadevan [44], implemented a Bayesian model-based fatigue damage prognosis (FDP) based on the Bayesian-ARIMA model for the update process and real-time monitoring of surface cracking in a rotorcraft. The acceleration and mass data measured from strain gauges were used to predict the sequence of loading. Kosorus et al. [45] present the usage of a TS identification model such as ARIMA and assess the usage of data mining method in the SHM system.

1.2.4 Autoregressive model with exogenous input (ARX)

The ARX is a nonlinear TS model that is popular for damage detection of structures because of its simplicity of application. Besides, the use of this model considerably reduces the computation costs since its application is feasible by having the data from only a single sensor. The estimation procedure operated by ARX is based on the polynomial estimation techniques which provide an efficient approach neglecting the disturbances in the analyzed signal. Bernal et al. [46] examined the relation between output error (OE) or equation error (EE) residual in autoregressive with exogenous input (ARX) models to identify the damage of an FE aluminum beam structure. Consequently, Monte Carlo analysis and experimental results concluded the robustness of OE residual compared to EE one in detecting the damage. Yang et al. [47] developed fault detection schemes through F-test, ARX, and time-synchronously averaged (TSA) techniques. The calculated residual signal from the ARX model is an indicator of the presence and location of damage in a gearbox under different load excitation. Oh Et al. [48] proposed a combination of support vector machine (SVM), AR-ARX, and nonlinear principal component analysis model to damage diagnosis of an eight-DOF mass-spring model under a time-varying excitation. The experimental results demonstrate the effectiveness of the proposed methodology in damage detection, data normalization, and speed of operation. Gul and Catbas [49] present a damage detection methodology based on the extracted coefficients of applying the ARX model on the various sensor clusters as the damage features of three case studies including a 4 degree of freedom numerical model, an international benchmark, and an experimental steel grid structure model to consider the noisy conditions. Furthermore, Sohn et al. [50] combined AR and ARX techniques to identify the structural changes of a surface-effect fast patrol boat model in conjunction with the Mahalanobis distance-based method. Different conditions of the boat assessed the robustness of the aforementioned techniques. Li and Ren [51], Optimized the energy consumption of an SHM system using PCA and AR-ARX model integrated with the genetic algorithm used to optimize the clustering algorithm. The usage of PCA is to decrease the dimension and size of data then AR-ARX is applied to extract the damage-sensitive factor.

There exist several studies in the literature utilizing the combination of AR and ARX models for damage detection of structures. Fasel et al. [52] evaluated the performance of ARX combined with Extreme Value Statistics (EVS) as the extracted features to nonlinear damage detection in a three-story building model. Lakshmi and Rao [53], proposed a robust output-only damage detection method using the AR-ARX model to indicate the time instant of damage in conjunction with probability density functions (PDFs) to design a DI. The result from the experimental benchmark three-story frame demonstrated the robustness of the proposed methodology. Krishnasamy et al. [54], improved the sensitivity of a DI using AR-ARX and augmented the singular spectrum analysis (SSA) model by predicting the errors. Experimental results from a simply supported RC beam validate the efficiency of the enhanced methodology. A damage detection approach of the composite beam model is investigated by Poulimenos and Salellariou [55] using the ARX model and transmissibility concept. Besides, Ugalde et al. [56], monitored an eight-story shear building model using the vector ARX model to divide the case study into multi-DOF substructures through acquiring the displacement data. Furthermore, the application of the AR-ARX model is examined in the structural damage diagnosis system by Silva et al. [57]. Damage indices are calculated based on residual error and quantified by a fuzzy c-means clustering method. Tatsis et al. [58] implemented the ARX model to detect the change of factors in the deficiency of structural stiffness under an environmental change in conjunction with the Gaussian process to improve the detectability and localization of damage.

1.2.5 Autoregressive moving average with exogenous excitation (ARMAX)

One of the serious drawbacks of the ARX model is that this model is suitable for the analysis of vibration data only with a constant signal-to-noise ratio (SNR). However, the data from the experimental models and full-scale structures in the real world do not have constant SNRs. This problem was solved by employing the autoregressive moving average with exogenous input (ARMAX) model in which the disturbances are considered to be included in the analysis of dynamic vibration responses of structures. The main advantage of the ARMAX model is to efficiently analysis of noisy signals and estimate the unbiased parameters of data with nonlinearity. Besides, it can efficiently provide all capabilities of the AR, ARX, and ARMA models. The basics of operating this model are on developing mathematical descriptions for dynamic structural responses with nonlinearity. A nonlinear model in fault identification in reinforced epoxy plates was proposed by Wei et al. [59] using the nonlinear ARMAX model. Damage indices are defined based on variation in modal features. Lakshmi et al. [60] enhanced the sensitivity of a feature extraction model to define damage indices using SSA and ARMAX techniques to locate the minor crack of a simply supported RCC beam model experimental case study. Do and Gul [61] implemented a robust and novel damage identification system using adaptive zero-phase component analysis (AZCA) and ARMAX techniques to automatically detect the change of mass and stiffness of a four-story shear laboratory model. Experimental results proved the accuracy of the presented methodology in determining the existence, severity, and location of the damage. Lakshmi and Rao [62] proposed a new damage detection system using ARMAX of a numerical simply supported beam model, eight-DOF benchmark, and finite element model (FEM) 12-DOF steel frame data set. Cepstral distance is used as a damage indicator between ARMAX models under healthy and damaged states. A new substructural damage identify system was designed by Mei et al. [63] based on optimal subpattern assignment (OSPA) and ARMAX techniques to quantify and locate the damage of an experimental three-dimensional RC model. Damage indices are defined based on the OSPA distance of an ARMAX model function. Xie and Mita [64] used Kolmogorov-Smirnov (K-S) to evaluate the differences of residual of ARMAX model as a damage identifier to indicate the presence of damage of ten-story shear numerical model and five-story experimental structure. Gislason and Gul [65], presented an automated SHM damage detection system based on ARMAX on a two-case study model including a four-story steel FE model and reinforced concrete buildings numerical model. Saaed and Nikolakopoulos [66] implemented a damage detection of a steel frame model by ARMAX as a linear parametric model. Xing and Mita [67] proposed a parallel and distributed methodology in damage detection of a shear five-story structure using the ARMAX technique. Experimental and simulation results reveal the applicability of the developed methodology.

1.2.6 Vector autoregressive

The ARV model is commonly used as one of the nonlinear TS models. This model has advantages in identifying the dynamic characteristics of the vibration responses without considering the signal disturbances generated during the dynamic excitation phase or due to environmental variations. This model characterizes the structural damages with high accuracy by using specific parameters such as the predictive errors, and the damage-sensitive features in which the damage intensity is predicted based on the calculations of the differences between the measured and modal data (i.e., residual errors). Li et al. [68], proposed a method based on modal identification to extract the energy proportion index of vibration from the vertical cantilever beam through the ARV model. Damage indices are calculated based on the energy index by estimating the modal parameters. Mosavi et al. [69] employed a damage localization and detection methodology using Vector AR (VAR) models of a two-span continuous steel beam subject to ambient vibrations. Lakshmi and Rao [70] defined a damage diagnostic model based on the combination of Dynamic Quantum Particle Swarm Optimization (DQPSO) algorithm, ARX, and ARV models to damage detection, location, and classification of a simply supported beam model. Mattson and Pandit [71] designed a method based on ARV models by extracting modal parameters such as mode shape, damping ratio, and shifting frequencies as damage-sensitive features of experimental and analytical data sets. Mattson and Pandit [72] introduced a novel damage detecting and locating approach based on ARV to identify the outlier prediction error of the experimental benchmark laboratory model. Okasha et al. [73] proposed an SHM system for statistical damage detection and reliability analysis of naval craft models by applying the ARV model on strain data sets. Bodeux and Golinval [74] utilized autoregressive moving average vector (ARMAV) models to identify the damage in the 'Steel-Quake' benchmark test structure as shown in Figure 1.4. Mosavi et al. [75] compared two damage detection methodologies based on using ARX and ARV of a simply supported steel beam model. Lakshmi and Rao [76] proposed an effective SHM model for damage detection purposes of two benchmark ASCE models using AR-ARX and ARV to calculate the damage indicator to identify and locate single or multiple damages. Kraemer and Fritzen [77] implemented an SHM system to monitor the dynamic characteristics of the offshore wind energy plants (OWEP M5000–2) by ARV to extract the modal parameters including mode shapes and eigenfrequencies as damage indicators.

1.3 FREQUENCY-DOMAIN SIGNAL PROCESSING TECHNIQUES

One of the most important factors in the determination of the dynamic responses and health monitoring of structures is frequency. This is because of the significant dependency of key structural characteristics of structures on this factor. As a well-known rule, the occurrence of damage in a structure would result in a reduction in its natural frequencies. In SHM, the measurement of the frequency parameter not only reduces the costs of experimental

Figure 1.4 View of the 'Steel-Quake' structure [74].

measurements but also its acquisition and analysis are facile. Also, the high accuracy of frequency measurements can be more pronounced when experimental data are obtained under controlled circumstances [78]. The damage detection process based on the variation of natural frequency has been commonly utilized in SHM [79] in which the shifting ratio of two frequency modes is considered as an indicator. The basics of frequency-domain techniques aim to evaluate the behaviors of the modal parameters of structural responses such as natural frequencies, modal shapes, damping ratio, etc. The frequency-based damage detection approaches deem any changes in the modal parameters to be changed in the structural characteristics including mass and stiffness. These techniques can be categorized into four types in terms of employed modal parameters including techniques (i) based on the frequency changes, (ii) based on the changes of the mode shapes, (iii) based on the occurrence of changes in the damping ratio, (iv) based on updating the structural model parameter. The main advantages of the techniques working based on the frequency changes compared to those based on the mode shape parameters are that these techniques need only the data from a single sensor location for estimating the structural responses and detecting damage. Besides, the techniques based on the damping ratio changes and updating model parameters are not popular and reliable in damage detection due to their complexity of applications and calculations. A damage detection approach based on the variations of mode shapes and frequency of structures was proposed by Stubbs and Osegueda [80]. A similar approach was adopted by Doebling et al. [81] to damage identification of structures. The shifts in the natural frequencies of a single-story framed structure were also observed by Nikolakopoulos et al. [82] to identify the severity and location of structural damages. A review study was carried out by Salawu [83] on the existing damage detection techniques based on the frequency shifting property. The importance of the frequency and modal features of structures was recognized as a key factor in the SHM of structures. A damage detection approach based on frequency changes was adopted by Williams and Messina [84] to identify the existence and location of damage in beam components. Morassi [85] demonstrated the successful performance of a frequency-based technique in identifying the damage in a vibrating rod. SHM of various types of structures and infrastructures such as high-rise buildings, bridges, and other laboratory-scaled structural components has been widely conducted in the literature by assessing the natural frequencies of the structures. Nagayama et al. [86] successfully performed the SHM of bridges based on the observation of the changes of natural frequencies. Also, the SHM of an RC bridge based on the frequency changes was carried out by Soh et al. [87]. Table 1.2 presents a summary of previous studies concerning the application of prominent frequency-based signal processing techniques including Discrete Fourier transform (DFT), Frequency Response Function (FRF), Strain Frequency Response Function (SFRF), FDD, Multiple signal classification (MUSIC) for SHM of various structures.

Aktan et al. [88] reviewed the SHM of various structures using modal analysis techniques and revealed that conducting dynamic tests is essential to investigate vibration-based problems. The SHM of Geumdang Bridge in Korea was performed using a frequency-based method by Lynch et al. [89]. In this method, the structural responses of the bridge were evaluated using a modal analysis technique. Comanducci et al. [90] conducted the SHM of suspension bridges using frequency changes due to dynamic effects of winds on bridges. The SHM of a long-span bridge was accomplished by Yarnold and Moon [91] using a numerical-based technique and analyzing the mode shapes and natural frequency features of the bridge. Besides, a finite element-based approach was presented by Zong et al. [92] for SHM of a full-scale continuous prestressed girder bridge using its frequency changes. Also, the existence of damage in railway axles was investigated by Rolek et al. [93] using the assessment of its vibrational responses in low-frequency ranges.

Table 1.2 Summary of using time-domain signal processing techniques in damage detection of civil structures

Study	*Technique*	*Analyzed structures*	*Level of damage detection*
Cheraghi et al. [96]	DFT	A pipe (L)	Existence
Lee and Kim [97]	DFT	A bridge (L)	Existence, location, and severity
Amezquita-Sanchez et al. [98]	DFT	3D truss-type structure (N)	Existence
El-Shafie et al. [106]	DFT	An IASC-ASCE benchmark model (L)	Existence, and severity
Philibert et al. [107]	DFT	A T-joint model reinforced by carbon fiber polymer (CFRP) (R)	Existence
Lynch et al. [108]	DFT	A five-story scaled model (L)	Existence and location
Zhang et al. [109]	DFT	250-m super-tall building (R)	Existence, location, and severity
Alsaadi et al. [111]	FRF	A composite plate (L)	Existence and location
Zenzen et al. [112]	FRF	A truss structure (L)	Existence
Lin and Loh [113]	FRF	A six-story steel frame (L)	Existence, and severity
Pu et al. [114]	FRF	A RC beam (L)	Existence
González et al. [131]	FRF	A two-story RC frame	Existence and location
Fallahian et al. [139]	FRF	A two-story frame (N)	Existence and severity
Zhang et al. [155]	SFRF	Steel stringer bridge scaled model (L)	Existence
Lee and Eun [154]	SFRF	A beam model (L)	Existence and location
Cheng and Cigada [156]	SFRF	beam-like structure (N, L)	Existence
Shadan et al. [157]	SFRF	FE truss frame model (N)	Existence, location, and severity
Wu et al. [164]	FDD	an RC beam bridge model (L)	Existence and location
Makki Alamdari et al. [165]	FDD	RCC beam (R) and two-fixed-end beam (L)	Existence, location, and severity
Jiang and Adeli [177]	MUSIC	A 38-story RC building (R)	Existence, location, and severity

Note: N, numerical; L, Laboratory; R, real-life.

1.3.1 Discrete Fourier transform

DFT is one of the primitive and widely utilized signal processing methods in the SHM area which is calculated based on an algorithm named Fast Fourier Transform (FFT). DFT can be performed simply and quickly to convert the stationary discrete-time sample signals into the frequency domain. DFT has been used in many various types of structures in real and laboratory environments to identify the damage by extracting the modal parameters because of its simplicity of application and computational efficiency. However, it has serious drawbacks in analyzing the responses of real-world structures due to their nonlinearity and nonstationary properties. This is because of the deficiency of the DFT algorithm in identifying the variations of the spectral behavior of the analyzed data over time (i.e., time-frequency domain) which is vital in the SHM of structures. Besides, DFT is not able to adequately capture the original information of signal data having closely spaced modes and noisy signals. Despite considerable disadvantages of the DFT algorithm, this technique has been widely used in damage detection of structures and some of the most prominent research works associated with the application of DFT in SHM, are reviewed as follows. Brincker et al. [94], Yuen and Katafygiotis [95] used DFT to identify the structural modal parameter of three different case studies including a 3D two-story frame, a 2D ten-story frame, and a seven-story RC building called Mong Wai under ambient noise excitations. Cheraghi et al. [96] employed the DFT method to identify the presence of damage in the pipes model structure. Lee and Kim [97] utilized DFT to detect the damage in a scaled steel bridge model under impact loading. Amezquita-Sanchez et al. [98] applied the DFT method to identify the modal parameters of a truss-type scale model exposed to earthquakes and forced excitation. Different applicability of the DFT in structural damage identification was previously presented [99–103]. Deficiency of the DFT algorithm in counter to the monitoring of the nonlinear and nonstationary response of real structures is the most important limitation of DFT. In addition, FFT cannot illustrate the changes of natural frequencies over time [104,105]. El-Shafie et al. [106] proposed a new high-resolution spectral analysis for structural damage detection approach using FFT and Fast Orthogonal Search (FOS) techniques. An IASC-ASCE benchmark model is utilized to conclude the advantages and limitations of the designed technique. Philibert et al. [107] identified an SHM system using DFT in damage identification of a carbon fiber-reinforced polymer (CFRP) T-joint model under impact load.

The combination of DFT and AR TS model is used by Lynch et al. [108] to develop a damage diagnosis model as a structural monitoring approach of a five-story scaled model as a case study. To enhance the usage of DFT in damage identification methodology Zhang et al. [109] employed the fast Bayesian DFT technique as a modal analysis preformation of a 250-m super-tall building model case study. Change of modal parameters is defined as a damage indicator in different conditions of the structure. Loewke et al. [110] investigated an SHM model of smart composite material based on two-dimensional DFT to produce a monitoring system through micro sensors and low-power networks.

1.3.2 Frequency response function

FRF is a linear-frequency-based transfer function that consists of real and imaginary numbers to calculate magnitudes and phases. Each function is a spectral function of a Fourier transform to compute the relationship between a response of the structure and an applied excitation. The main advantage of FRF is associated with its capability in analyzing complex structures that demonstrate noisy vibration responses with high SNRs. Besides, FRF data can provide more effective information about the vibration response in comparison with limited information captured by modal data such as coordinates and number of modes. However, the main disadvantage of the FRF method is that it cannot efficiently obtain important information about damage type, size, and location. Alsaadi et al. [111] designed a damage identification and location methodology using FRF and least-squares rational function to determine the dynamic response of a composite plates model through Micro-Electro-Mechanical System (MEMS) sensors. The usage of a combination of FRF, Genetic Algorithm (GA), and Bat Algorithm (BA) in damage identification methodology of truss structure was examined by Zenzen et al. [112]. The results illustrate the improvement of accuracy in finding the location and classifying the severity of damage through the proposed approach. Lin and Loh [113] investigated the use of FRF and Sammon's mapping as a dimensional reduction algorithm to consider all the acquired sensor data of the two case studies including a six-story steel frame and twin tower structure as shown in Figure 1.5. An optimization damage quantification based on FRF is presented by Pu et al. [114]. In this study, an RC beam model is under study as an experimental case study to investigate the application of damage detection methodology.

Some research studies used the FRF as a damage indicator such as Zang et al. [115] and Sampaio et al. [116] which applied the correlation criterion between intact and damage FRFs as a damage detection index. Sampaio et al. [117] and Mondal et al. [118] utilized an FRF mode shape curvature as the damage indicator. Reddy and Swarnamani [119] presented the usage of FRF curvature energy to detect a plate structure's damage. Although Kim et al. [120] and Shi et al. [121] demonstrated that the FRF curvature depends on the accuracy of the central difference approximation, there is a need for a significant number of measuring locations for adequate resolution. One of the principal defects of the FRF-related damage indicators is the difficulty to

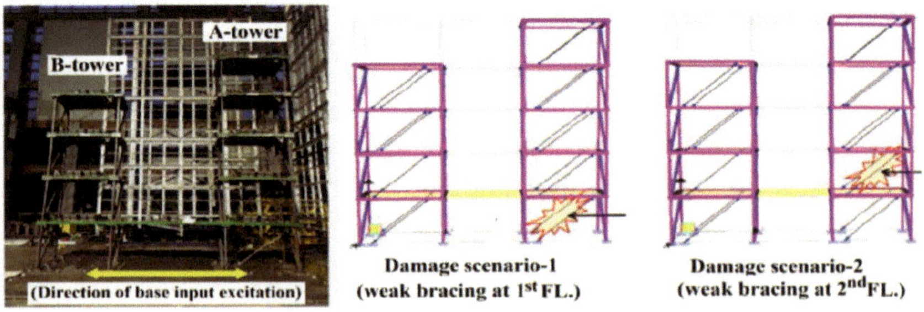

Figure 1.5 Schematic diagram of the test specimen and its damage scenario [113].

connect the design parameter and damage location of the structure to classify damage severity [122]. One of the popular damage identification approaches is based on the use of the FRF technique in combination with the FEM of various systems. This technique is efficiently able to identify the presence of structural damage and detect its location by evaluating the variations of the FEM parameters. The successful use of this method for damage identification of different systems has been reported by several previous works [123–132]. In line with the study on the FRF-based damage detection approach used by Lin and Ewins [123] for a framed building structure, Wang et al. [125] validated the applications of this method for updated models. Esfandiari et al. [126] utilized the FRF data including the natural frequencies and modal shapes to estimate the location and severity of damage in a beam member.

Although the aforementioned model updating techniques require adopting a series of sensitivity equations, the obtaining procedure of these equations is complex since the data of all degrees of freedom need to be measured. Hence, the need for data approximation techniques increases to enhance the feasibility of this procedure. However, adopting data approximation techniques such as the FRF data expansion method would come along with a set of unwanted errors and uncertainties in the analyzed data. Besides, the implementation of analytical approaches as the data estimation methods requires high accuracy of the analyzed FEM and the measured data [127]. Therefore, several research works [128–130] focused to solve these drawbacks by transforming the basic theory of the FRF from matrix level to the DOF level which can lead to enhance the feasibility of the damage detection process by direct comparison between the FRF data in damaged and undamaged states. González et al. [131] proposed a damage identification methodology based on FRF to estimate the dynamic properties of a two-story RC frame model as shown in Figure 1.6. The ambient and forced vibration signals of the analyzed case study are used to extract the frequency values. Castro et al. [132] designed an impedance-based damage detection methodology based on FRF and conventional electro-mechanical impedance (EMI) method under low SNR condition of an aluminum plate scaled model.

Most studies used FRF as input of ANN as a successful part of damage detection methodology. A combination of the FRF technique and ANN was used by Padil et al. [133] to capture the variations of frequency-domain parameters of a steel truss bridge with simple supports. As another example study on the combination of the FRF and ANN, the FRF data of a two-story framed structure was considered as the inputs of ANN by Dackermann et al. [134] to automate the FRF technique for different scenarios varying in terms of the mass and connection pattern of the structural members. It was found that employing the hybrid approach could efficiently enhance the capability of the damage localization process by optimizing the selection procedure of sensor data for the application of the FRF. The combined applications of the FRF and PCA techniques were also employed by Nguyen et al. [135] as the damage identification of a model of Sydney Harbor Bridge modeled by

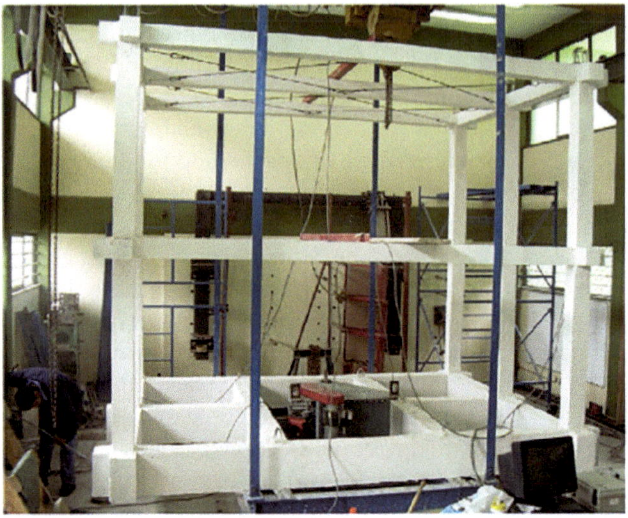

Figure 1.6 Testing pendulum with the RC frame used for the identification of structural damage [130].

a concrete arch beam. In the proposed method, a specific feature (DSF) was developed to be considered as the input of the ANN. The capability of this technique in recognizing the severity of the damage was concluded.

Another hybrid approach based on the combination of the FRF and two-dimensional PCA (2D-PCA) was proposed by Khoshnoudian et al. [136,137] to reduce the FRF data size. The acceleration FRF (AFRF) data obtained from the numerical model of a cantilevered composite sandwich beam were analyzed by Wang and Zhao [138] with eliminating the FRF data uncertainties to identify the structural damages. Furthermore, Fallahian et al. [139] proposed a structural damage identification using FRF and Mode-shapes as the input of Couple Sparse Coding (CSC). A numerical two-story frame model is established to verify the aforementioned damage detection approach and demonstrates the effectiveness of CSC and FRF in damage detection and severity. Saaed and Nikolakopoulos [66] used the ARMAX model for the construction of FRF to identify the natural frequency as a damage indicator of the ten-story benchmark steel frame model. The reliability of the hybrid method as a robust structural damage identification technique was revealed.

1.3.3 Strain frequency response function

The estimation of the axial or flexural strain responses as a strength parameter of infrastructure and structure can reflect the local features which used to update the dynamic model. Using this parameter in the frequency domain can be considered as FRF data named SFRF. ||||In the last decade, some research studies have taken advantage of SFRF to analyze a structural dynamic. Early research on estimating the dynamic parameters of the FE model was introduced by Ip and Vickery [140] using the SFRF technique to calculate the stresses of the structure. Moreover, SFRF and the least square method were utilized as a sensitive feature in an analytical dynamic of a structural model presented by Esfandiari et al. [141]. Their findings indicated that all analytical mode shapes need to estimate the SFRF. Guo et al. [142] introduced an updating of a structural dynamic model using SFRF on numerical and experimental preformation that demonstrated the robustness of the proposed methodology. Besides, using SFRF in damage detection research works are reviewed as follows. Cornwell et al. [143] reported an early work based on using strain energy parameters as a damage indicator in damage identification of the numerical FE plate-like structure model. Sampaio et al. [116] extended the FRF based on displacement curvature technique to the SFRF curvature technique based on the difference between the baseline and damaged condition of SFRF values.

Concerning the damage-induced changes in a pair of strain and natural frequency shifts as dynamic sensitive parameters, a sensitivity study was investigated by Yam et al. [144]. Numerical and experimental setup of plate-like structures were evaluated to evaluate the robustness of the proposed damage identification methodology. Lee and Kim [145] proposed a damage identification algorithm using FRF and SFRF to extract the dynamic features of the experimental bridge model to feed up an input layer of the pattern recognition NN model. A modal-based damage detection methodology proposed by Park et al. [146] is investigated using the dynamic strain parameter. The experimental result demonstrates the feasibility of the method in detecting and locating the damage using fiber Bragg grating (FBG) sensors. Wu and Li [147] performed a two-level damage diagnosis methodology in flexural cantilevered beam structures based on using distributed fiber optic sensors through the extraction of dynamic strain parameters. Strain-based damage identification was investigated by Kesavan et al [148] as an SHM system for a 2D polymeric composite T-joint scaled model. Based on using the strain measurement, Katsikeros and Labeas [149] designed a structural damage detection based on modal analysis to identify the presence and location of a crack in aircraft cracked lap-joint model. Li [150] presented a review study of different strain-based techniques for detecting the damage and evaluating the feasibility and advantages of the SFRF method. The result of the comparison study indicated the better preformation of SFRF due to the independent to numerical modal extraction. A damage identification using FBG in conjunction with strain feature and SFRF as a DI based on the differences of SFRF between before and after damage scenarios. [151]. Regarding acquiring dynamic strain as a damage indicator in damage assessment Loutas et al. [152] investigate an intelligent SHM system based on FBG optical sensors in a composite structure model. A damage detection procedure based on strain and accelerometer sensors is designed by Lee et al. [153] using SFRF and displacement frequency response function (DFRF) together. Moreover, structural damage identification of a beam model using DFRF and SFRF techniques was introduced by Lee and Eun [154]. Zhang et al. [155] provided a structural damage identification based on strain modal processing through the SFRF technique to indicate the damage indices of a steel stringer bridge scaled model. Application of strain modal parameter as a significant indicator in SHM purpose is assessed by Cheng and Cigada [156] using complex mode indicator function (CMIF). A beam-like structure is used as an experimental and numerical case study to assess the performance of the proposed methodology. Shadan et al. [157] calculated damage-sensitive parameters through the SFRF to structural damage localization, detection, and quantification of a FE truss frame model.

1.3.4 Frequency domain decomposition

The FDD technique is known as a popular frequency-domain algorithm for operational modal analysis (OMA) which was proposed by Brincker et al. [158] to overcome the existing deficiencies of basic frequency domain (BFD) or Peak-Picking (PP) method in recognizing the modal parameters of modes that are closely spaced each other. In this method, the singular value decomposition (SVD) based on the matrix of output PSD is used to estimate the modal parameters of the output-only systems. Although the identification of the modal parameters of structures using traditional modal analyses by adopting transfer functions is feasible through the analysis of extracted data from the exciting structure, the application of artificial excitations to large structures and infrastructures would be complicated. Also, the SNR of signals from large-scale structures under ambient excitations in the real world would be considered due to the noisy environments which can affect the frequency responses of the structure. Accordingly, this can cause difficulties in characterizing the structural damages due to the occurrence of unwanted overlaps in the signal frequency band. Hence, conducting controlled-condition experimental tests on laboratory-scale structures has been raised in recent decades that lead to low SNRs. To overcome the difficulties arising from the SHM of large structures with noisy response signals, the FDD and the enhanced frequency domain decomposition method (EFDD) techniques were proposed by Brincker et al. [158,159], respectively to enhance the frequency band separability and consequently identification of modal characteristics of extracted from noisy signals [160–163]. Wu et al. [164] employed a damage identification approach of an RC bridge model using the FDD technique in conjunction with strain mode decomposition. The experimental results demonstrate the robust performance of using FDD to identify the strain mode parameters in a noisy environment. In the proposed method, FDD was firstly applied to fuse data from multiple sensors and extract damage-sensitive features of a steel structure. A random projection algorithm is then used to reduce the dimensionality of the data followed by a self-tunings OCSVM for anomaly detection. Makki Alamdari et al. [165] proposed a damage assessment of two experimental case study including the RCC beam from the iconic Sydney Harbor Bridge and two-fixed-end beam as shown in Figures 1.7 and 1.8, respectively. Abdelgahani et al. [166] investigated the performance of INOPMA under realistic excitation and we compare it to the FDD algorithm. An example is a 3DOF system under El Centro seismic excitation. It is shown that both methods perform equally well.

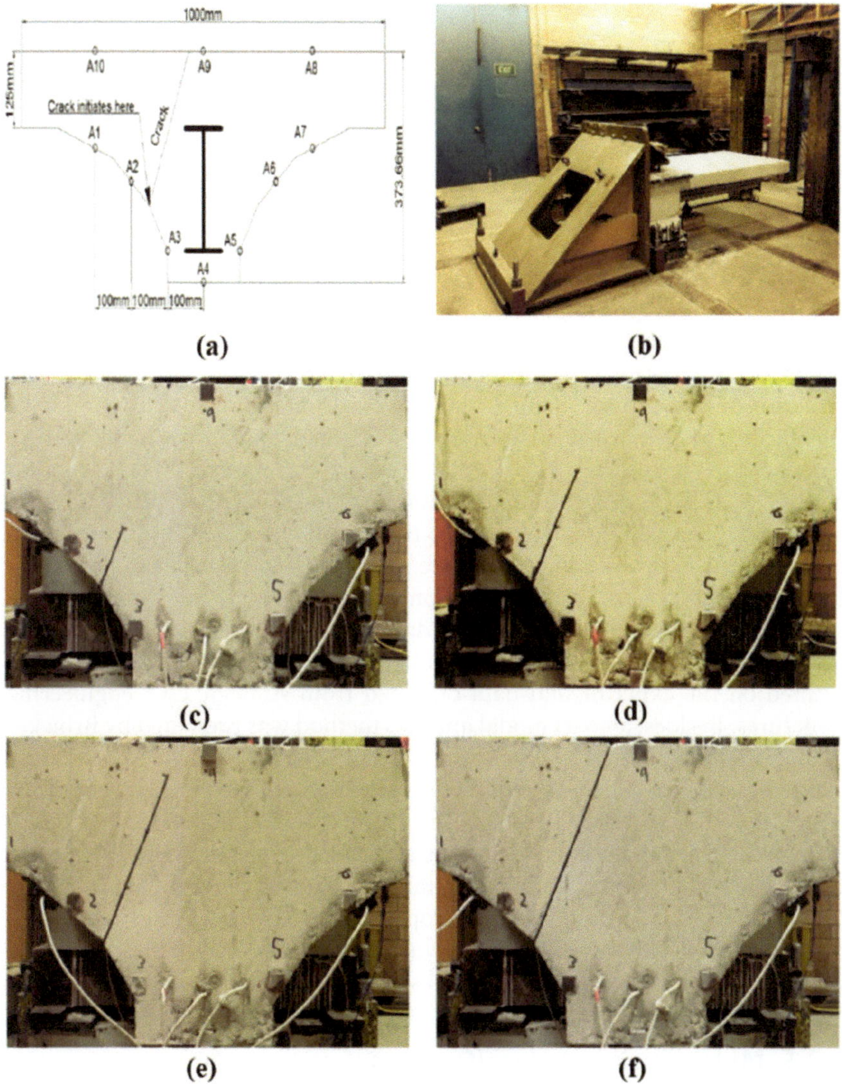

Figure 1.7 The crack introduced into the test specimen: (a) Structure with an arrow indicating the cut, (b) support of the structure, (c) Damage Case 1: 75-mm damage cut, (d) Damage Case 2: 150-mm damage cut, (e) Damage Case 3: 225-mm damage cut, and (f) Damage Case 4: 270-mm damage cut [165].

(a) (b)

Figure 1.8 Illustration of (a) the bus lane on the Sydney Harbor Bridge and (b) one of the concrete jack arches underneath the bus lane [165].

An improved FDD technique was used by Liu and Venkatasubramanian [167] for oscillation monitoring of power systems by investigating phasor measurement units (PMUs). Also, a novel technique based on FDD was proposed by Pioldi et al. [168] to explore the modal characteristics of framed structures when exposed to different seismic ground motions. By employing this technique, the feasible variations of common modal characteristics of the structure were investigated under various earthquakes. The combined actions of FFD and the Global Positioning System (GPS) technique were employed by Gorski [169] to evaluate the modal properties of experimental displacement data including the mode shapes, damping ratios, and natural frequencies of structures in comparison with those obtained from FE analyses. The performances of various modal analyses based on the output-only technique were investigated by Magalhaes and Cunha [170] to be implemented on the experimental data obtained from strategic civil engineering structures. Besides, a novel modal analysis method was presented by Brincker et al. [171] in which the FDD technique was adopted to identify the modal properties of output-only systems in comparison with the performance of the conventional PP method. The capability of this method in the decomposition of high-frequency and noisy data was concluded. The short time-frequency domain decomposition (STFDD) technique was used by Malekjafarian and Obrien [172] to identify the modal properties of the vibration responses of a bridge under the action of moving vehicles. The proposed method was efficiently able to produce a global modal shape vector of the whole bridge.

1.3.5 Multiple signal classification

To improve the detectability of close frequencies in comparison with the performances of traditional techniques such as FFT in analyzing noisy signals with low SNRs, the MUSIC algorithm was proposed which provides a high-resolution power density spectrum [173]. However, the process of the

MUSIC technique needs a large amount of computational time leading to the increase of computation costs. The capability of MUSIC in the direction-of-arrival (DOA) estimation [174] and the damage detection of induction motors [175] was concluded. By employing a nonparametric system identification model developed by Adeli and Jiang [176], a novel method was presented by Jiang and Adeli [177] as a real-time and non-destructive SHM approach based on the combination of neural network and MUSIC algorithm for damage detection of a 38-story RC framed structure and the efficiency of the proposed technique in the damage identification of high-rise building structures was concluded. The effective use of the MUSIC algorithm was employed by Osornio-Rios et al. [178] for SHM of a truss under forced excitations to detect the structural faults due to internal corrosion.

To improve the representation of data by MUSIC algorithm from frequency-domain to time-frequency representation, the short time MUSIC (ST-MUSIC) technique was proposed by Garcia-Perez et al. [179] in which the MUSIC algorithm is applied during the short time intervals with different spectrums. The ST-MUSIC is efficiently able to that provide a robust nonlinear analysis of stationary data even with transient characteristics. The combination of MUSIC and artificial neural network algorithms named MUSIC-ANN was presented by Rios and Sanchez [180] to detect the damage location and classify its severity in a five-bay truss structure in which the amplitude and the natural frequencies were adopted as the input data. Figure 1.9 illustrates that the methodology of the proposed approach was an effective simple, automated, and reliable tool in damage detection in SHM. The capabilities of this method in fault diagnosis of induction motors [181] and determining the natural frequencies of a spatial truss structure [182] in noisy environments have been demonstrated in the literature. This is proved the efficient performance of this technique in analyzing data with closely spaced modes and natural frequencies [183].

As the applications of the MUSIC algorithm in mechanical systems, a MUSIC-based damage detection approach was proposed by Perez et al. [182] for fault diagnosis in an induction motor. Also, a hybrid method from the combination of the EMD and MUSIC methods was used by Martinez

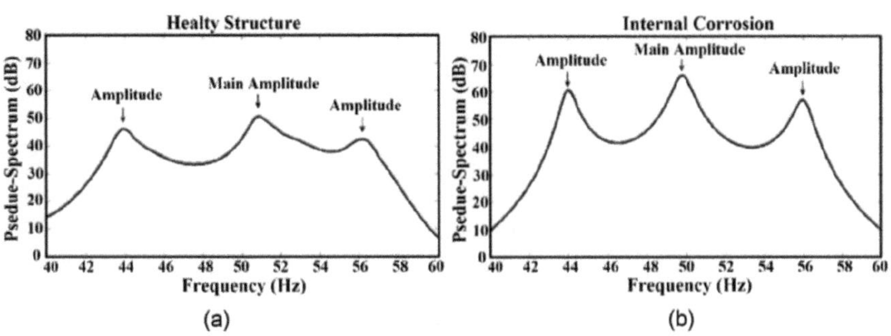

Figure 1.9 Pseudospectrum of the synthetic signal [178].

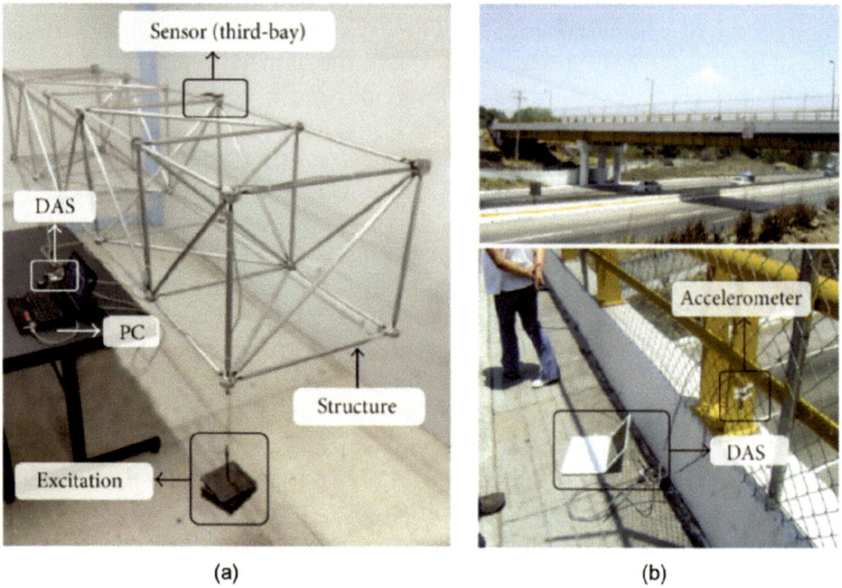

Figure 1.10 Experimental set up of; (a) a scaled model of a truss structure and (b) a pedestrian bridge [184].

et al. [183] to characterize the existing faults in the frequency responses of induction motors. In line with this study, Martinez et al. [184] developed a novel hybrid approach from the combination of ensemble empirical mode decomposition (EEMD) and MUSIC techniques named the EEMD-MUSIC method to determine the natural frequencies and the mode shapes of a scaled model of a truss structure and a pedestrian bridge as illustrated in Figure 1.10a and b, respectively. In this method, the vibration signals were firstly decomposed by the EEMD to a set of intrinsic mode functions (IMFs), then the natural frequencies of the decomposed data were identified by using the MUSIC. In addition, the reliability of the proposed approach was examined using different synthetic data. To identify the damages in a composite plate with carbon fiber material commonly used in the oil tank of aircraft, Zhong et al. [185] used a 2D-MUSIC method based on a spatial smoothing algorithm. Similarly, Zuo et al. [186] employed a 2D-MUSIC technique for damage identification of the numerical-based data of aluminum plate and experimental signals of a laminated composite plate structure. A MUSIC-based damage identification approach was proposed by Perez-Ramirez et al. [187] to detect and locate the damages in a truss structure exposed to different dynamic excitations. This method was operated based on variations of the signal amplitudes and the measurement of pseudo-spectra. Gkoktsi and Giaralis [188] introduced a new MUSIC-based method to determine the natural frequencies of a 3-story framed structure under white noise excitations by adopting the simple PP technique.

A modified MUSIC-based algorithm was proposed by Bao et al. [189] to solve the existing phase errors in various types of sensors and improve the localization capability of the MUSIC algorithm. The reliability and accuracy of the proposed method in the damage localization were revealed through its application for a composite panel reinforced by a T-shaped stiffener. As an extension of 2D-MUSIC algorithms, Elbouchikhi et al. [190] developed a multi-dimensional MUSIC (MD MUSIC) algorithm for fault diagnosis of bearing systems through the identification of the natural frequencies. Bao et al. [191] presented a novel MUSIC-based approach to improve the application of the MUSIC algorithm for damage identification of complex structures as illustrated in Figure 1.11. Based on this method, the fault diagnosis and damage localization procedures were improved by solving the existing phase delay disorders in the signals.

A MUSIC-based damage detection method was proposed by Fu et al. [192] for SHM of a structure using two arrays of piezoelectric sensors under various excitations. The capability of the proposed method was also evaluated by motoring a circular structure bonded using a metal ring. Yuan et al. [193] improved the damage localization performance of a MUSIC-based technique in the velocity phase of the signal data of a composite aircraft

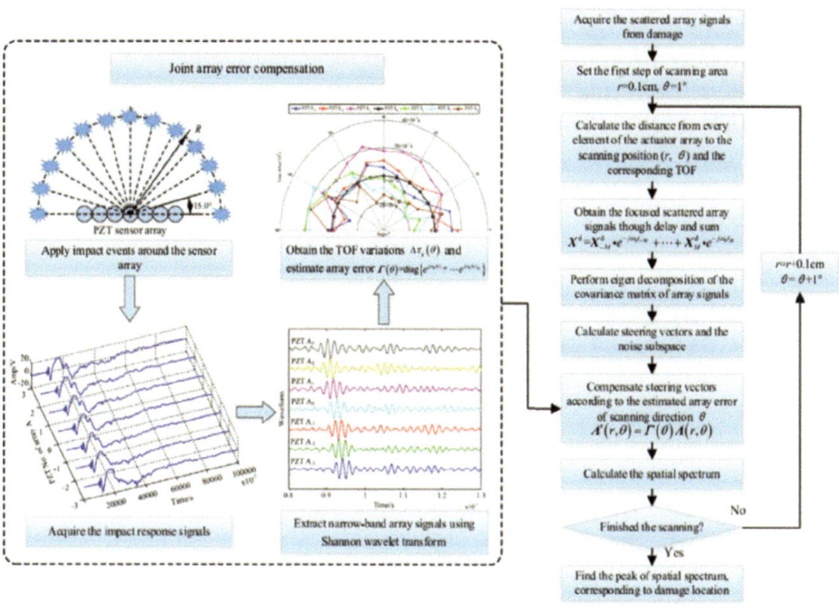

Figure 1.11 Schematic diagram of the new synthesis aperture-MUSIC algorithm array error compensation [191].

wing box reinforced by a T-shaped stiffener. Based on the assumption of the guided wave propagation in near fields, a new modal-based 2D-MUSIC algorithm was employed by Zou et al [187] as a damage detection approach of the numerical model of an aluminum plate and the experimental model of a laminated composite plate. The damage localization performance of a time-reversal multiple signal classification (TR-MUSIC) method was evaluated by Dimopoulos et al. [194] in composite structures using a series of independent scatters. It was found that this technique is effectively able to estimate the location of the damage by reducing the number of required sensors and enhancing the accuracy of data classification. Also, damage imaging using time-reversal imaging with MUSIC technique was proposed by He and Yuan [195] in a metallic plate and its efficiency was revealed as a successful signal processing technique.

This section presents a comprehensive review of the vibration-based signal processing techniques for dynamic feature extraction of structures in the time and frequency domains that have been commonly used to detect and locate structural damage in various civil engineering structures when subjected to different excitations under environmental or controlled conditions by Zhang et al. [196]. Different time- and frequency-dependent techniques were categorized based on their feature extraction procedures and applications. Furthermore, the applications of signal processing techniques and their ability level in damage identification of different structures were reviewed. From the review on time-domain techniques, it is found that these techniques are mostly used in linear systems while they are highly sensitive to noise. Among the time-domain techniques, ARMAX has more advantages in analyzing nonlinear and nonstationary signals compared to other models and it efficiently can mitigate the noise disturbance of vibration responses of structures. A summary of the advantages and disadvantages of time-domain and frequency-domain signal processing techniques is illustrated in Figure 1.12.

1.4 TIME-FREQUENCY-DOMAIN SIGNAL PROCESSING TECHNIQUES

With increasing urbanization and construction of civil engineering structures such as buildings, bridges, dams, etc. dependency on and needs for SHM of various structural systems are growing when subjected to natural disasters such as earthquakes, flooding, and other catastrophes or under weakening environmental factors leading to deterioration such as humidity, salinity, atmospheric acidity, scour, and wind. The analysis of nonlinear and nonstationary signal responses of different structural [197–201] and mechanical [201–206], systems especially real-life structures exposed to various dynamic loads and noisy environments always challenges traditional signal processing techniques with time-domain or frequency-domain representations for

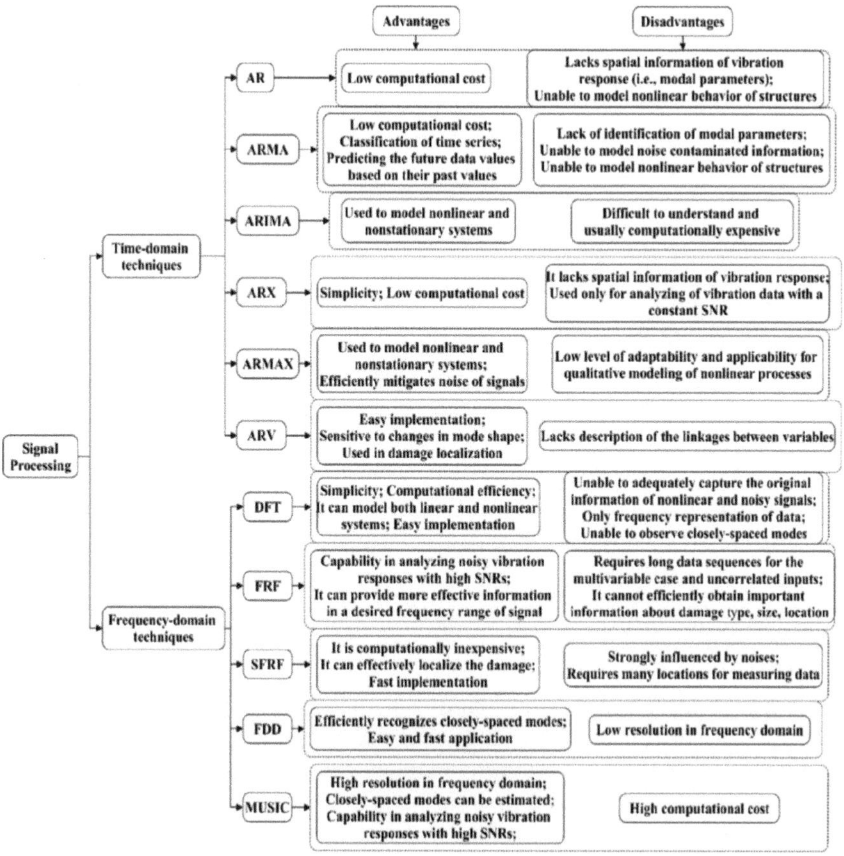

Figure 1.12 A brief summary of advantages and disadvantages associated with time-domain and frequency-domain signal processing techniques [196].

damage detection, localization, and quantification purposes. Traditional signal processing techniques have serious deficiencies in the analysis of nonlinear and noisy signals that contain closely spaced frequencies. Therefore, utilizing robust signal processing techniques to capture the nonlinear properties of the vibration response of structures is of paramount importance in damage localization, quantification, and detection process [207]. Hence, many attempts have been made in recent decades to improve the performance of signal processing techniques by proposing time-frequency analysis approaches that are efficiently able to capture the fundamental information of analyzed signaled with high-resolution representations. The successful performance of time-frequency signal processing approaches has been extensively reported by previous studies, especially for SHM of real-life structures representing noisy signals with nonlinear characteristics. These techniques are even efficiently able to identify invisible damages that occur in the internal section of the structures through the variation of nonlinear structural characteristics. Despite existing several review studies in the literature on vibration-based structural damage detection approaches, there exists no review in categorizing the applications of signal processing techniques with time-frequency representations. This study presents a comprehensive review on the development of time-frequency signal processing techniques in recent decades. The applications of time-frequency analysis techniques in various scientific fields of aerospace, civil, and mechanical engineering are reviewed by clarifying their advantages and deficiencies.

1.4.1 Short time Fourier transform

To overcome the drawbacks of the FFT technique as a traditional signal processing technique in the presentation of spectral time-variations, an extension technique named short time Fourier transform (STFT) was introduced by Gabor [1] that can efficiently represent the frequency contents in the time domain. STFT is known as an alternative and improved FFT technique that can effectively analyze nonstationary signals. That is, these techniques is able to capture the local frequencies that occur during very short time periods in the response vibrations that cannot be recognized using FFT. The procedure of recognizing the instant local frequencies over the time using the STFT is based on dividing the signal time-history response into short time intervals (i.e., time windows) each of which is analyzed using the FFT in frequency domain during the specified short time period of each window [208]. The STFT is mainly employed to identify the time-frequency representation of the vibration responses of structures due to its capability to present more understandable information along with useful definitions of the time-frequency distributions of the analyzed signals. The STFT technique has been widely used in the literature for identifying the modal parameters of different structures such as reinforced concrete (RC) frames [209], steel frames [210], truss-type structures [174], and beam members [98]. One of the main shortcomings of the STFT is related to the dependency of its accuracy and the resolution of its results on the analyzed window size. As such, to capture high-resolution frequency results and identify the information of closed natural frequencies, the time window size should be sufficiently large [211]. However, the behavioral trend of transient signals is estimated through the analysis of large-size time windows efficiently. Figures 1.13a and b illustrate examples of the time-frequency distributions of the acceleration record of El Centro earthquake (1940, N–S) and analyzed vibration responses of a seven-story RC framed building under this earthquake obtained in terms of the daily length of day data [210].

Yesilyurt and Gursoy [212] employed the STFT technique to identify the modal damping ratios of a composite beam. However, the modulus information of this member including Young's modulus in fiber direction,

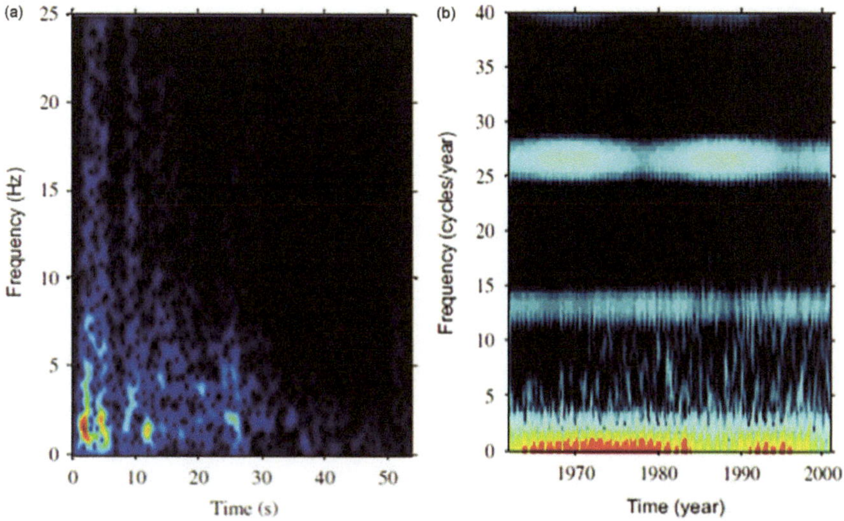

Figure 1.13 (a) Time varying spectra of the acceleration record of El Centro (1940, N–S) and, (b) the daily length of day data of a seven-story RC framed building obtained by the STFT method [210].

shear modulus, and transverse Young's modulus was characterized using frequency-domain analyzes. In addition, an analytical approach was proposed based on the use of the Hanning window to capture the modal shape and damping information of a three-degree-of-freedom system by analyzing the vibration signals during their free vibration phases. Also, the successful application of the STFT was revealed in identifying the vibration features of the signals compared to the estimation performance of a validated Q-factor technique.

Dolce and Cardone [213] demonstrated the accurate performance of STFT in representing the spectral features of the signal responses of shape memory alloys (SMAs) simultaneously in both time and frequency domains. A DI based on the averaged STFT spectrogram method was adopted by Cocconcelli et al. [214] for damage detection of direct-drive motors. The proposed DI was calculated based on the marginal integration in time series of the signal responses. The performance of the STFT in the implementation of an output-only damage identification method on multi-degree of freedom (MDOF) systems representing linear time invariant (LTI) and linear time variant (LTV) behaviors, was evaluated by Nagarajaiah and Basu [215] in comparison with the performances of empirical mode decomposition (EMD) wavelet techniques. It was found that the STFT can efficiently characterize the damage as well as the EMD and wavelet methods. The successful application of the STFT has been also

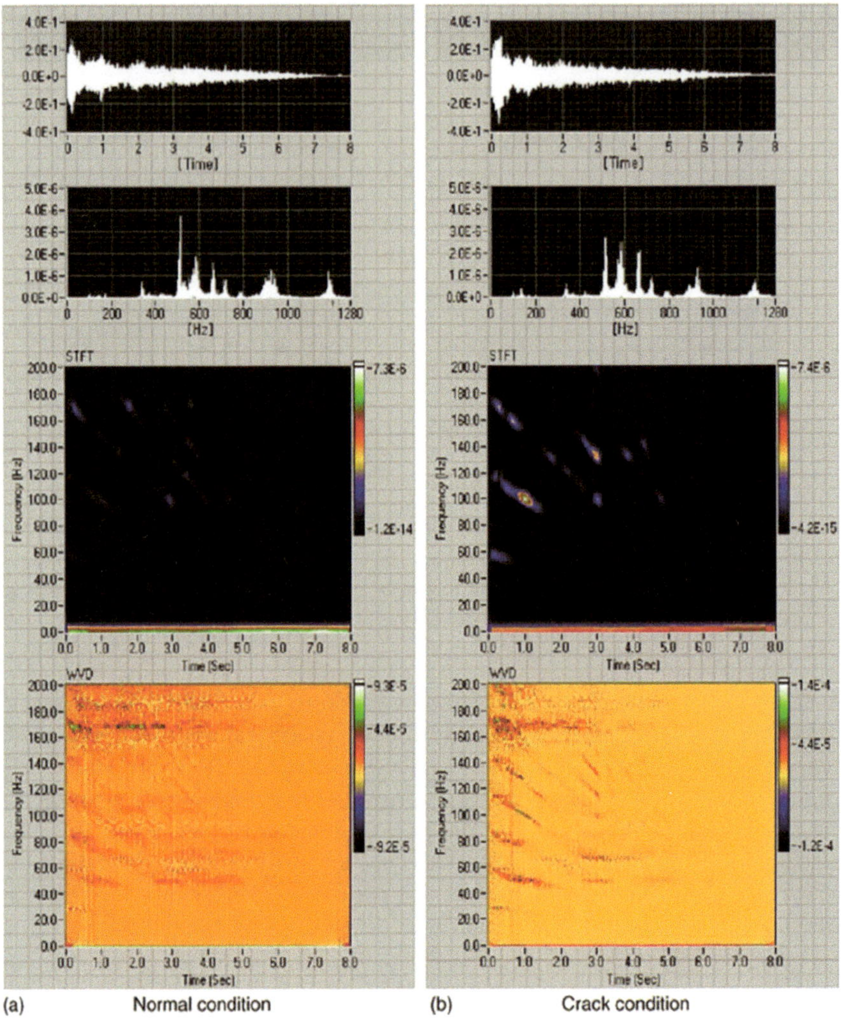

Figure 1.14 Time signal, FFT, STFT, and WVD of (a) normal and (b) crack condition during acceleration [222].

reported in SHM of tuned mass dampers [216–221] and mechanical systems [98] as well as the capabilities if the END and Hilbert Transform (HT) techniques. Kim et al. [222] carried out a comparative study on the feature extraction performances of various time-frequency techniques including the FFT, STFT, Wigner–Ville Distribution (WVD), and discrete wavelet transform (DWT) in damage detection of a spindle-typed rotor-bearing system. As shown in Figure 1.14, the STFT methods can identify the time and frequency characteristics of the damage as well as the WVD and present better performance compared to FFT. A time-frequency algorithm

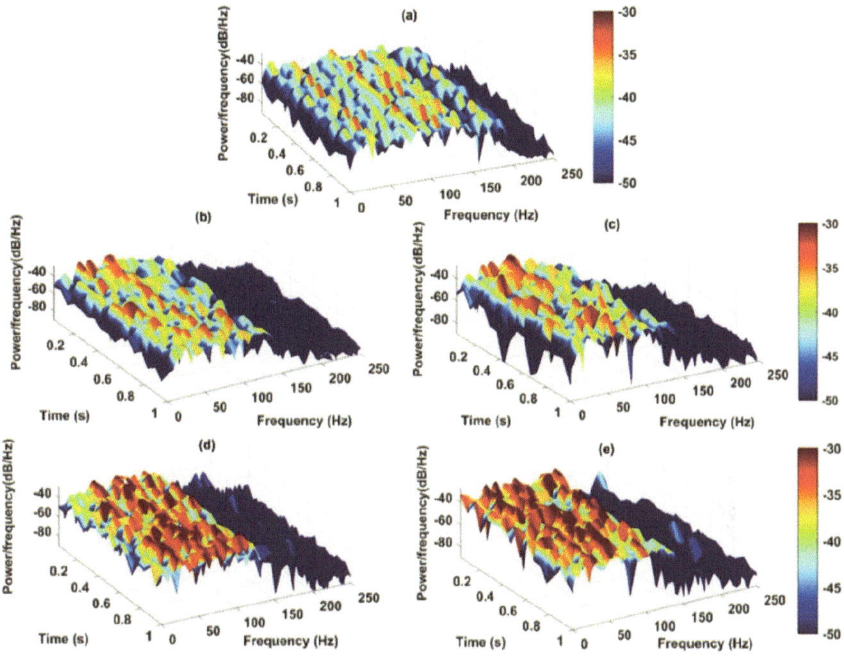

Figure 1.15 Spectrograms of the IMFs of sensor-10 under five different states of the truss including; (a) healthy, (b) 20% damage, (c) 50% damage, (d) 80% damage, and (e) 100% damage [224].

based on the application of the STFT was proposed by Fitzgerald et al. [223] for damage identification in the numerical model of the blade of a wind turbine by analyzing the rotations velocity and the stiffness features of the blade. The sensitivity of the proposed method was concluded to the existing of structural faults due to the stiffness changes and frequency changes varying in time distributions.

The capability of the STFT technique in the damage detection, localization, and classification of a laboratory-scale steel truss bridge model was demonstrated by Mousavi et al. [224]. As such, STFT was utilized to illustrate the change of frequency content when a series of damages accrued in a steel truss bridge model. As shown in Figure 1.15, the magnitude of instantaneous frequency (IF) power enhances in the low-frequency ranges compared to the healthy state as the damage severity increases.

1.4.2 Hilbert Huang transform

Frequency-based signal processing techniques have substantial drawbacks in deriving the time-related features of the analyzed signal [225]. Also, these techniques cannot properly determine the information of higher frequency modes that basically represent the characteristics of the signal faults [226–228]. To solve these issues, several improved signal processing techniques such as wavelet transform (WT) [229–231], Hilbert–Huang transform (HHT) have been proposed in recent years. HHT is a novel signal processing technique that was proposed by Huang et al. [232–234] based on the combination of HT and EMD techniques to analyze non-linear and nonstationary signals. The procedure of EMD is based on the decomposition of signal data into a series of finite IMFs as such it can efficiently capture the time-dependent instantaneous responses of the analyzed signal. The EMD has been widely used in the literature for the damage detection of civil engineering structures representing nonstationary and nonlinear responses in the real world [235–240]. However, there still exist substantial issues with the implementation of HHT for SHM of structures. To overcome the exiting drawbacks of the EMD such as mode mixing problem, EEMD, and complete ensemble empirical mode decomposition with adaptive noise (CEEMDAN) techniques were proposed as the improved HHT techniques that will be reviewed in the following sections. Table 1.3 presents a list of some indexed research works utilizing the HHT techniques on different analytical, numerical, and experimental models of various types of structures with representations in time-frequency domains.

Table 1.3 Summary of using frequency-time domain signal processing techniques in damage detection of civil structures

Authors	*Name of technique*	*Analyzed structures*	*Level of damage detection*
Rezaei and Taheri [240]	EMD	Cantilevered steel beam (N, L)	Existence
Dong et al. [241]	EMD	The Imperial County Services Building (R)	Existence
Cheng-Zhong and Lian Xu-Wei [242]	EMD	Transmission tower (R)	Existence
Xu and Chen [243]	EMD	Three-story shear building (L)	Existence
Sarmadi et al. [244]	EEMD	Four-story steel frame of a IASC-ASCE model (L)	Existence and location
Entezami and Shariatmadar [245]	EEMD	Four-story steel structure (L)	Existence and location
Mousavi et al. [224]	CEEMDAN	Steel truss bridge (L)	Existence, severity, and location
Fakih et al. [246]	CEEMDAN	Friction stir welded joints (N, L)	Existence and severity

Note: N, numerical; L, Laboratory; R, real-life.

1.4.2.1 Empirical mode decomposition

The basic procedure of EMD is adopting a sifting process to decompose the measured signals into a set of IMFs through and consequently extracting the feature information in time and frequency domains. In this technique, the characteristics of at least two extrema in terms of their magnitudes and time intervals are required for decomposing the raw signals. The main advantage of EMD compared to frequency-based techniques is that this technique can efficiently analyze the nonlinear and nonstationary signals and represent the information of the extracted features in both time and frequency domains with relatively high resolution. There exist many research works in the literature that have reported the application of the EMD in analyzing the vibration signals recorded from civil engineering structures. The capability of the EMD technique was proved by Pines and Salvino [247] in damage detection of experimental and analytical models of one-dimensional (1D) structures. Rezaei and Taheri [240] proposed a novel damage indicator based on the energy feature of the vibration signals extracted by EMD to detect the size, severity and location of notch damages in the numerical model of a cantilevered steel beam. The HHT based on EMD method was employed by Chen et al. [248] for SHM of the Guangzhou New TV Tower (GNTVT) as shown in Figure 1.16 during the construction and service phases of the tower when subjected to typhoons

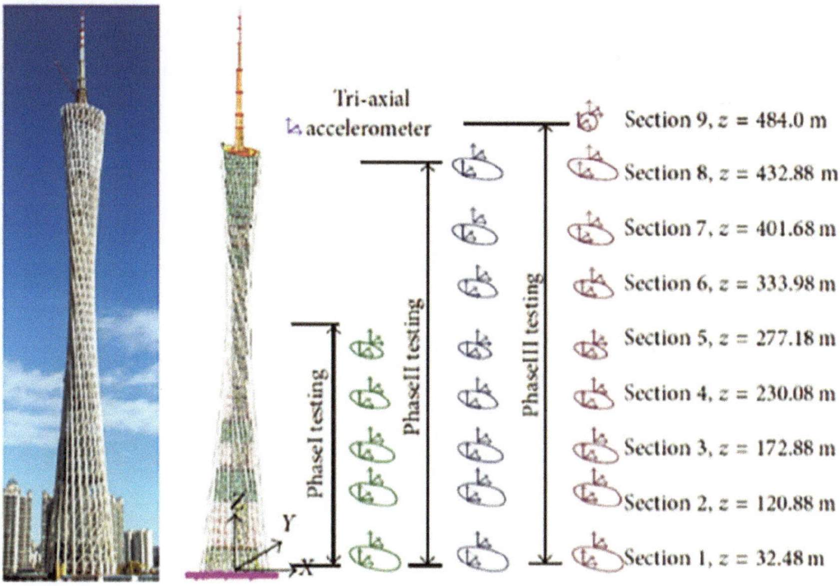

Figure 1.16 Guangzhou New TV Tower [248].

and earthquakes. Once acquiring the signal data from different sections of the tower as shown in Figure 1.16, the HHT was applied to signal modes decomposed by EMD for identifying the frequency-based features such as the natural frequencies and damping ratios, and time-related parameters including the instantaneous frequencies and energy features represented in time-frequency domains compared to the results from WT techniques.

A hybrid damage detection approach from the combinations of the EMD and a TS-based feature extraction technique named vector autoregressive moving average (VARMA) model was proposed by Dong et al. [241] for SHM of the Imperial County Services Building and the Van Nuys Hotel. The procedure of the proposed method was based on the detection of sudden energy variations in time-frequency distributions especially when occurring at high ranges of the frequency responses. The combination of EMD-HHT techniques was used by Cheng-Zhong and Lian Xu-Wei [242] for damage identification of a transmission tower by characterizing the shape factor feature of the analyzed signals. Xu and Chen [243,249] employed the EMD for the damage detection of a laboratory model of a three-story shear building under vibration excitations applied through a shake table as shown in Figure 1.17. An application of an adaptive HHT method for identifying the modal responses of a composite beam was reported by Bao et al. [250]. It the proposed method, the HHT was improved by using an autocorrelation algorithm combined with a band-pass filter for the purpose of noise reduction in the measured signals. The EMD was used by Rezaei and Taheri [240] for damage detection and classification in the numerical and experimental models of a steel beam. In this study, the energy feature of the first IMF obtained from the decomposition process by EMD was considered as the damage indicator.

As a field application of the EMD in the real world, He et al. [251] used the combination of the EMD and the random decrement technique (RDT) for identifying the modal characteristics of a railway bridge in China as

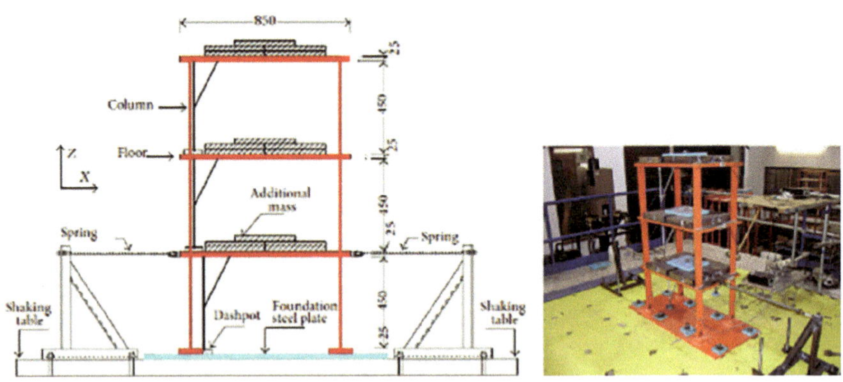

Figure 1.17 Configuration of the building model (unit: mm) [248].

Figure 1.18 A photograph of the NYR steel truss bridge [251].

shown in Figure 1.18. In addition, an FE-based comparative study was carried out to evaluate the capability of the EMD-RDT method compared to the results from performing the PP approach. The superiority of the EMD-based technique was concluded in modal identification of the studied bridge. An application of EMD-based HHT was also reported by Bao et al. [250] for structural damage identification, and detecting the damage location and its severity when subjected to ambient excitations.

Pavlopoulou et al. [252] conduced the SHM of an aluminum panel and fiber-epoxy composite laminates through the analysis of ultrasonic waves guided on these components using an EMD-based HHT technique. As another application of the EMD-based HHT method, Ghazali et al. [253] employed this technique for the damage detection in experimental and analytical models of a pipeline system. Esmaeel and Taheri [254] demonstrated the successful performance of a damage identification approach based on the application of EMD on the vibration signal data of a high-rise building structure. Also, the sensitivity of the proposed method to the damage severity and its dimensions was revealed. Alvanitopoulos et al. [255] also concluded the capability of the EMD-based HHT technique in identifying and characterizing the structural damage in RC-framed structures through the analysis of the extracted signal features including peak and average values of the signal amplitude parameter. An efficient performance of a DI based on the instantaneous frequency feature extracted by the EMD-based HHT method from the vibration responses of soil-quay wall systems was concluded by Wei et al. [256] in detecting the seismic damages in these systems. In addition, this DI was successful in identifying the incidence of liquefaction and the time-dependent characteristics of the soil. The combination of EMD and RDT techniques was employed by Shi et al. [257] for characterizing the modal parameters of the Shanghai World Financial Center structure including the natural frequencies and damping ratio parameters when exposed to different excitations. Although the combined EMD-RDT technique accurately could estimate the natural frequencies of the structure, it was failed in identifying the damping ratios when the structure was exposed to ambient dynamic excitations. A combination of wavelet packet transform (WPT) and EMD techniques was used by Garcia-Perez et al. [258] for damage detection, localization, and classification of its severity in a steel truss by defining an energy-based DI. This hybrid method was effectively able to detect the presence and location of the fault due to various factors such as the loosened bolt at the joint connections, cross-sectional stiffness reduction, and corrosion.

Chiou et al. [259] introduced a new damaged indicator defined based on the variations of the damping ratio feature of the vibration responses analyzed by the EMD-based HHT technique for damage identification of various benchmark models when exposed to different earthquake excitations. As an improvement on the traditional acoustic emission (AE) techniques, Lin and Chu [260] used the damage indices defined based on the energy and instantaneous frequency features extracted from the acoustic emission data in characterizing the damage at the joint connections of a steel offshore structure under tensile stresses. Similarly, an improved AE technique using EMD-based HHT was employed by Hamdi et al. [261] for damage identification of composite structures through a series of bending tests. The capability of this technique in the analysis of the nonstationary AE signals was concluded. The validation of the EMD-based HHT method was also examined by Lin et al. [262] in analyzing the impact-echo signals with high noise for damage detection of concrete components. Yadav et al. [263] used the EMD for extracting the signal features of the ultrasonic wave data obtained from a steel channel under moving loads to characterize the damage presence in this structure. Damage detection based on the application of EMD on the acceleration responses of a beam member was utilized by Meredith et al. [264]. The capability of this method in classifying even minor cracks with a severity of 10% under moving loads, was concluded. Roveri and Carcaterra [265] also revealed the efficiency of the EMD-based HHT technique in damage detecting and localization of bridges using the analysis of signal data measured from only a single point under moving traffic loads. Also, Yu and Ren [266] concluded an accurate performance of EMD in identifying the modal characteristics of the Xining Beichuan Arch Bridge as shown in Figure 1.19.

The performance of EMD-based HHT in analyzing the nonlinear and nonstationary buffeting responses of a long-span bridge was evaluated by Ma et al. [267]. The EMD technique captured more accurate results in the time-frequency domains compared to FFT in identifying the frequency spectrum of the responses. A combination of EMD and spectral analysis was used by Chen et al. [268] for SHM of the Tsing Ma Suspension Bridge by measuring the strain, temperature, displacement, and acceleration responses data. The proposed hybrid technique was efficiently capable to capture the static and dynamic features of the measured noisy signal responses.

Figure 1.19 Xining Beichuan Arch Bridge [271].

Zhang et al. [269] employed the EMD for analysis of the vibration responses of a bridge system under wind excitations to identify the modal parameters including the instantaneous frequencies and damping ratios. The EMD could successfully detect the structural faults by realizing the behavioral changes in the modal characteristics. A hybrid damage identification technique from a combination of EMD and wavelet method was proposed by Yi et al. [270] to recognize the dynamic features of the wind responses of a high-rise building under typhoons. Despite the capability of the EMD in capturing the vital dynamic and time-dependent features of signal, there exists a considerable mode mixing drawback associated with the performance of the EMD in analyzing signals having very close or the same frequencies [271].

1.4.2.2 Ensemble empirical mode decomposition

To overcome the mode mixing problem of EMD, the EEMD technique was introduced by Wu and Huang [271] in which a white Gaussian noise is separately added to each IMF of analyzed signal with an identical standard deviation. The application of the EEMD has been widely reported in the literature for the purpose of solving the mode mixing issue of the EMD. A combination of EEMD and RDT methods was used by Liu et al. [272] in which the statistical autoregressive vector (ARV) was employed as the feature extraction tool for characterizing the modal parameters of the Xing Nan Bridge under ambient dynamic excitations. It was found that although the EEMD could not only efficiently identify the modal features of signals without any mode mixing problem, it requires more computational costs compared to the use of EMD because of utilizing the added white noise in obtaining the IMFs [273]. By comparing the performances of the EMD and EEMD techniques in analyzing the seismic signals by Wang et al. [274], the superiority of the EEMD in the decomposition of the IMFs without any mode mixing problem and better presentation of the results in time-frequency domains was concluded. Besides, Lin [275] found that a combined use of EMD and EEMD has some advantages compared to the sole application of the EMD in the analysis of the responses of a bridge. A hybrid signal processing technique based on the combined EEMD and the MUSIC technique was proposed by Martinez [184] to characterize the modal characteristics of a truss bridge model. The superior performance of the proposed method in the presentation of the signal features time-frequency domains was concluded compared to the performances of traditional techniques such as DWT and FFT methods. Amiri and Darvishan [276] carried out a comparative study on the performances of the EEMD and EMD techniques by adopting a density-based clustering technique in evaluating the acceleration signals of a steel frame based on the analysis of the frequency and amplitude features. An energy-based damage identifier proposed by Sarmadi et al. [244] was concluded by using EEMD as the feature extraction method combined with Pearson correlation function for damage detection and localization in an IASC-ASCE structure model when subjected to ambient vibrations as shown in Figure 1.20. The IASC-ASCE structure model was also used by Entezami and Shariatmadar [245] to

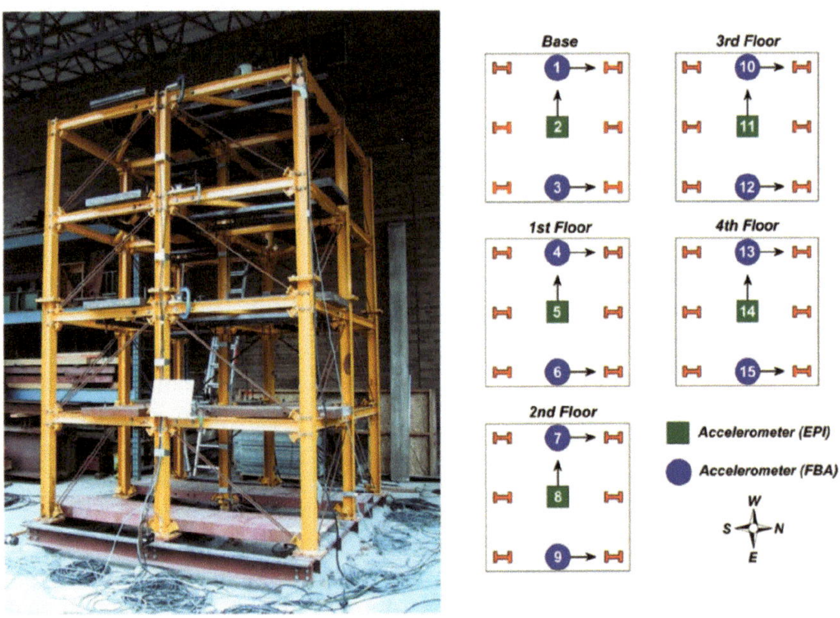

(a) (b)

Figure 1.20 (a) A four-story steel structure of studied as the IASC-ASCE structural health monitoring benchmark problem by Sarmadi et al. [244], (b) Sensor numbers and locations.

evaluate the performance of a new hybrid damage identification method from a combination of EEMD and auto regression model with exogenouinput (ARARX), and correlation-based dynamic time warping techniques. The capability of the proposed method was found in detecting the damage presence and its location.

1.4.2.3 Complete ensemble empirical mode decomposition with adaptive noise

Despite the aforementioned advantages of the EEMD over EMD, there exist two major drawbacks associated with the performance of EEMD due to adding white Gaussian noise to the signal leading to difficulties in obtaining the average of IMFs and existing residual noise in IMFs. To solve this issue, the CEEMDAN technique was proposed by Torres et al. [273] in which a particular noise is added to IMFs at each stage resulting in unique residues for each IMF. Additionally, CEEMDAN has a proper spectral separation which considerably reduces the time of decomposition process. The three aforementioned EMD-based techniques have been widely used in mechanical [277–280], geological, seismic signals [281], and electrocardiogram (ECG) [273] applications. However, CEEMDAN has been rarely used for SHM of civil engineering structures. The application of CEEMDAN in analyzing the seismic signals was reported by Han and Baan [282] to improve the resolution of the results in time and frequency domains and to solve the mode mixing issue of EMD. The performance of CEEMDAN in analyzing the acceleration responses of a steel-truss bridge model as shown in Figure 1.21 under white noise excitations was experimentally evaluated by Mousavi et al. [283] using four signal features extracted from the IMFs including the energy, IA, unwrapped phase, and

Figure 1.21 A perspective view of the truss in the laboratory of Qingdao University of Technology by Mousavi et al. [224].

instantaneous frequency. In addition, several damage indices based on the hybrid applications of two statistical time-history features, including kurtosis and entropy features with the energy and IA features were proposed. The superiority of CEEMDAN in detecting the presence, and location of damage and quantifying its severity was concluded compared to EMD and EEMD. Afterward, Mousavi et al. [224] utilized CEEMDAN in combination with ANNs for structural damage localization and quantification of a steel-truss bridge model experimentally subjected to white noise excitations as shown in Figure 1.21. Experimental results demonstrated the robustness and efficiency of using CEEMDAN in damage detection. In line with these studies, a combination of CEEMDAN and MUSIC technique named CEEMDAN-MUSIC was proposed by Mousavi et al. [284]. The results demonstrated the advantages of the proposed techniques in damage detection and localization compared to pure CEEMDAN.

Xiao et al [285] introduced an SHM model by employing CEEMDAN and HT to extract the spectrum signature of a bridge vibration under a vehicle moving load. Li et al. [286] introduced an energy-based feature extraction methodology using CEEMDAN for ship-radiated noise (S-RN). Lv et al. [287] applied CEEMDAN improved multivariate multiscale sample entropy (MMSE) to detect faults in a rolling bearing system and the capability of the proposed methodology was concluded. The capability of a fault diagnosis approach from a combined application of CEEMDAN and Adaptive Neuro-fuzzy Inference System (ANFIS) was reported by Kuai et al. [288]. A DI based on Improved CEEMDAN was introduced by Fakih et al. [244] to detect the presence and assess the severity of damage resulting from the flaws in friction stir welded joints. Experimental and finite element analysis reveals high sensitivity of the proposed method.

1.4.3 Wavelet Transform

Despite common applications of FFT in analyzing stationary signals that represent constant characteristics without nonlinearity, it is not efficiently capable to process nonlinear and nonstationary signals. The advantages of WT techniques in analyzing the nonlinear and nonstationary signals in time-frequency domains through adopting specific filters have motivated researchers to widely use these methods for SHM of real-world structures in recent years. The WT is also able to improve the drawbacks of the STFT in terms of the time-frequency representation of data. That is, the spectral characteristics of analyzed time series can be presented with higher resolutions by WT by performing a multiresolution analysis to identify the time-dependent variations in TS. The procedure of this resolution analysis by WT is based on the use of different-size wavelets adapted to the size of the target signal features. The use of the WT technique is not only limited to analyze the signals of civil engineering structures but also many research works have reported the expensive application of WT-based techniques in improved forms such as continuous wavelet transform (CWT), DWT, wavelet multi-resolution analysis (WMRA), WPT, synchrosqueezed wavelet transforms, empirical wavelet transform (EWT), etc. for SHM of various engineering systems [211,289].

1.4.3.1 Continuous wavelet transform

The CWT is a formal tool that provides an improved representation of a signal in time-frequency domains by adopting wavelets with different scales of parameters that continuously vary depending on the size of the target features. The signal faults commonly are detected using CWT based on the analysis, scaling, and transforming of the basic functions on a mother wavelet. Several previous studies have reported the use of CWT for the damage detection of different structural and mechanical systems.

Okafor and Dutta [290] employed CWT for damage identification in cantilever beams by determining the continuous wavelet coefficients through a Daubechies mother wavelet. The damage states were applied to the finite element (FE) models of the beams by reducing their stiffness. The CWT efficiently detected the damage location by characterizing the spikes in the signal behaviors. The Morlet mother wavelet was used in the CWT by Yoon et al. [291] for identifying the corrosion damage and its severity in RC beams through the analysis of AE signals. The CWT technique could successfully categorize the damage mechanisms and failure behaviors of the beams including localized minor cracks, flexural and shear damages by specifying their relevant frequency changes. The CWT based on processing the variations of ridge patterns was employed by Melhem and Kim [292] for detecting the presence of fatigue damage in RC slabs and beams with simple supports. As a result, it was found the number of clear ridges increased with increasing the damage severity and growing the cracks, especially in large-scale (with low-frequency responses) RC structures. In addition, the superiority of the CWT in analyzing the nonstationary signals was concluded in comparison with FFT. The validity of a hybrid damaged detection approach from the combination of the CWT by adopting the Morlet mother function and a two-step SVM classifier was demonstrated by Park et al. [293] for detecting and classifying the damages in railroad tracks as shown in Figures 1.22 and 1.23. The seismic responses of RC columns and a four-story steel frame were analyzed by Noh et al. [294] by using the CWT based on the Morlet mother wavelet for the damage detection. In this technique, the CWT adopted three damage-sensitive features (DSFs) to characterize the energy variations of the wavelets in proportion to changes in damage patterns of the structure. The reliability of the employed technique was approved by correlating the results to the classification of damage mechanisms using the drifts between the stories. Su et al. [295] proposed a hybrid damage identification approach from the combination of the CWT based on Daubechies mother wavelets and an efficient

Figure 1.22 Test specimen and built-in PZT patch [293].

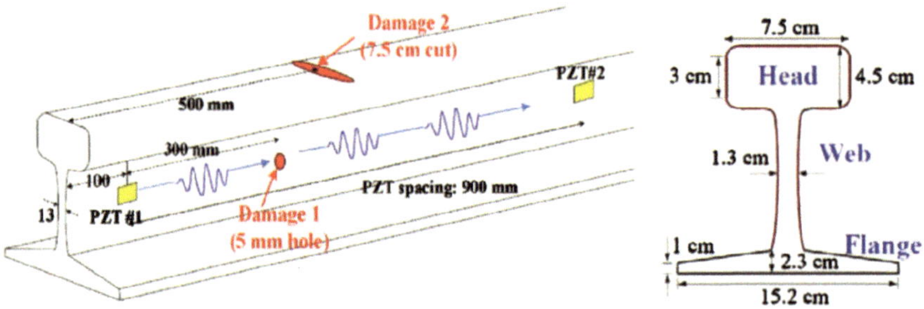

Figure 1.23 Damage descriptions on railroad section [293].

time-varying autoregressive with exogenous input (TVARX) to determine the instantaneous modal parameters of a linear time-varying structure under earthquake excitation. The proposed method was efficiently able to detect the location of the structural damage without any need for reference data. Li et al. [296] demonstrated that the application of a Butterworth filtering in combination with CWT adopting a Morlet mother wavelet could adequately detect the structural faults and crack patterns in a three-span cable-stayed bridge by removing the environmental effects.

A hybrid application of CWT and HT techniques was used by Gaviria and Montejo [297] for identifying the signal features and reconstruction of the analyzed signals of an RC structure under seismic excitations. The capability of the proposed method in obtaining the modal parameters including the natural frequencies and the damping ratio was concluded as well as identifying the structural characteristics including the mass and stiffness matrices. In addition, the capability of this technique was validated in efficiently detecting the presence, and location of the damage, and in classifying its severity. Gholizad and Safari [298] identified the structural damages in space structures through the analysis of the variations that occurred in the mode shape features of the signals by employing the CWT technique in which the Mexican hat mother wavelet was adopted. The existence and location of damages on the tubular elements and joints were adequately detected using the CWT method. With the intention of identifying the damage presence and location in a steel beam, Shahsavari et al. [299] proposed a CWT-based statistical analysis in which a Symlet wavelet was performed in combination with likelihood-based statistical methods. According to this method, first, the signal modes extracted by the application of the CWT were analyzed to detect the damage. Then, the analyzed signal modes were evaluated using a likelihood ratio test to detect the location of the damage. The difficulties associated with probabilistic analyses in computing the unbiased distribution of uncertainties were improved by utilizing a CWT-based non-probabilistic technique by Abdulkareem et al. [300] as a damage detection approach for a steel plate. The damage detection process of the method was based on a damage indicator defined by the determination of the upper and lower bounds of the wavelet coefficients. The CWT by adopting a Mexican hat mother wavelet was also utilized by Wang et al. [301] for damage identification and characterization in a metro tunnel. Also, a DI was defined based on the residual force vectors of the damaged structure compared to those of the healthy state of the tunnel. The capability of this method was concluded by detecting the damage location in the tunnel and classifying its severity. A hybrid damage detection framework by using kernel function based on the combination of SVM and CWT with a Morlet mother wavelet, HHT, Teager-Huang Transform (THT) was proposed by Pan et al. [302] for SHM of a cable-stayed bridge. Also, a parametric study was carried based with variations of the damage severity and its location. The superior performance of CWT in analyzing noisy signals was concluded compared to HHT and THT. As a real-world application, Karami-Mohammadi et al. [303] used the CWT for damage detection and localization in power transmission towers as shown in Figure 1.24 by determining the characteristics of their vibration responses. It was demonstrated that CWT was able to properly identify the structural faults by characterizing the modal parameters of the vibration responses such as the curvature of the mode shapes.

1.4.3.2 Discrete wavelet transform

The discrete wavelet transform (DWT) is known as another popular WT technique in signal processing which is performed by discretization of parameters in CWT. The main advantages of CWT are associated with the considerable capability of this technique in processing different scales and segments of signals, and the redundancy of CWT as such it represents significant overlap between wavelets at each scale and between scales. However, the application of CWT requires high computational cost since it represents a high-redundant data analysis approach. In contrast, DWT represents a non-redundant analysis requiring lower computational time due to the use of orthogonal functions in the decomposition signals. As much as extensive application of CWT for SHM of different systems, there exist

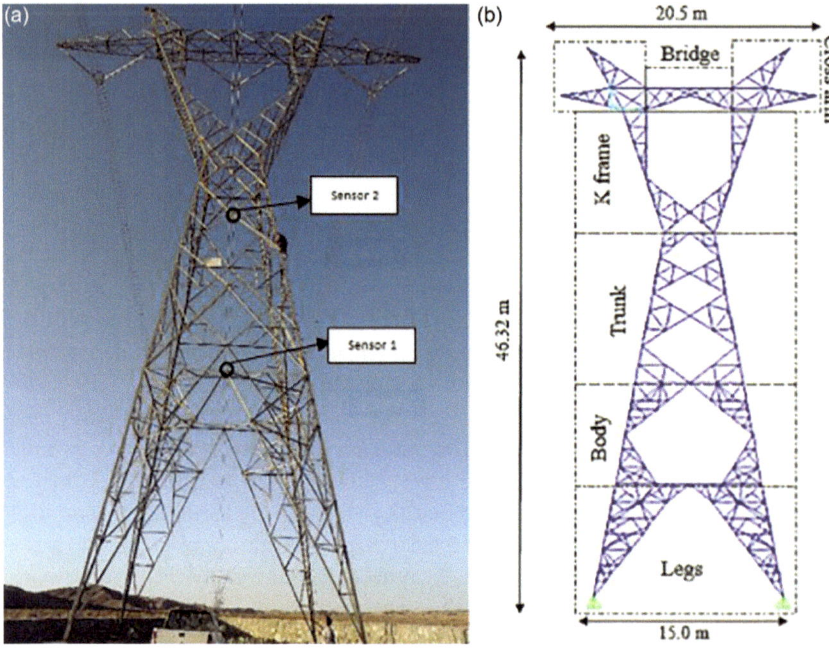

Figure 1.24 (a) Geometry and dimensions and (b) experimental test of the case study tower [303].

many research works in the literature reporting the widely use of DWT as a popular data processing technique with time-frequency representations. Hou et al. [304] employed CWT and DWT techniques by adopting the Daubechies mother wavelet for damage diagnosis in a simplified single-degree of freedom (SDOF) system. The capability of both WT-based techniques was demonstrated in detecting the existence and location of the damage through identifying the characteristics of the spikes observed in the wavelet responses of the signals. Hera and Hou [305] performed DWT for detecting the damage in a four-story steel frame structure of the ASCE benchmark model under wind load excitations. The DWT could efficiently identify the structural damages in the brace elements by characterizing the response spikes that appeared in the wavelets. Besides, the damage location was recognized using the distribution patterns of the spikes. Ovanesova and Sua´rez [306] successfully recognized the crack patterns, the damage severity, and its location in an RC beam using the application of DWT with different mother wavelet functions on the beam responses and consequently identifying the wavelet coefficients. The variations of the wavelet coefficients were identified by observing discontinuity behavioral trend and detecting any sudden changes in the resolution of the scalograms. Hoseini Vaez and Tabaei Aghaei [307] employed DWT for damage detect through observing spikes in the wavelet plots of the analyzed signal.

1.4.3.3 Wavelet multi-resolution analysis

The basic concept of WMRA is to perform DWT by adopting different filters. This technique is able to present different resolution data of an analyzed original signal including data with coarse and fine resolutions that contain information associated with low- and high-frequency signal modes, respectively. The application of WMRA in combination with DWT technique as a time-frequency technique has been reported in signal processing and SHM of various structural and mechanical systems by several previous studies [303,308–313].

A damage detection approach based on the use of coarse resolution analysis of WMRA to obtain low-frequency information of data from a series of beam members was employed by Feng et al. [308]. In this method, first, the strain data of the beams was analyzed using DWT and the wavelet coefficients of the decomposed signal were obtained. Thereafter, the fault features that were identified through additional transformations implemented on the captured WT coefficients using multiresolution analysis. The capability of the DWT-WMRA in detecting the location of cracks in the beams was concluded even in a noisy environment. As another hybrid signal processing technique utilizing WMRA, the successful performance of a hybrid damage identification approach from the combination of WMRA and ANN proposed by Lucero and Taha et al. [309] was demonstrated for an RC bridge. Furthermore, a combination of WMRA with the HT technique was employed by Djebala et al. [310] for fault diagnosis in a gear system. In this method, the signal modes decomposed using HT were further analyzed by WMRA to identify high-frequency fault features through high-resolution information. Datta et al. [311] also evaluated the performance of a hybrid technique from the combined applications of a WMRA and a neural network for diagnosing damages in industrial robots. The validity of the utilized approach was demonstrated by using the experimental data obtained from an industrial robot manipulator. As a practical study on the application of WMRA on structural members, Zhang et al. [312] analyzed the acceleration responses of a laboratory-scale model of an RC beam as shown in Figure 1.25 and the capability of a hybrid algorithm adopting WMRA in combination with a wavelet damage function was concluded.

1.4.3.4 Wavelet packet transform

Similar to the performance of WMRA, WPT which represents an extension form of WT, analyzes signals by classifying their components into high- and low-frequency modes. WPT is also able to completely decompose signals so that obtains signal time-frequency features representing both stationary and nonstationary characteristics [313].

This analysis method has advantages over WMRA in terms of providing a more detained and flexible analysis procedure for signals in which wavelet packets perform based on the linear combinations of wavelet functions.

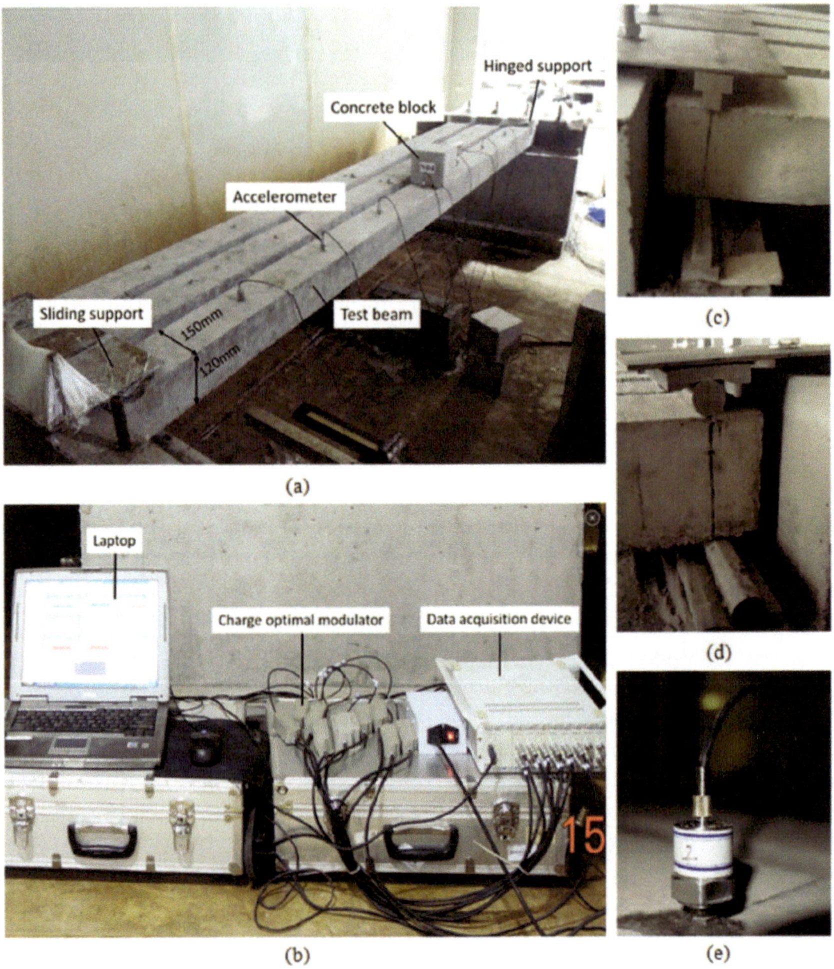

Figure 1.25 Experimental layout: (a) overview, (b) data acquisition instruments, (c) hinged support, (d) sliding support, and (e) piezoelectric accelerometer [312].

The application of WPT in damage detection and SHM of different systems has been extensively reported in the literature [314–320]. A hybrid application of WPT and the neural network techniques was used by Sun and Chang [314] for damage detection in a three-span continuous beam when subjected to impact loads. According to this method, the wavelet components obtained from the decomposition of the signals by WPT for different states of the beam were used as the inputs of the neural network. The proposed method was efficiently able to detect the presence, location, and severity of the damage. Yam et al. [315] used the energy-based damaged detection approach in which the energy feature of the vibration responses

Figure 1.26 Experimental model of a laboratory-scaled steel beam [320].

was obtained by using WPT. The capability of WPT in identifying the damage in steel beams was also presented by Han et al. [316] by using a novel damage indicator based on the wavelet packet rate. The existence of structural faults in the shear connectors of a reduced-scale model of a steel bridge was evaluated by Ren et al. [317] under impact loads based on the variations in the behavioral trend of the wavelet packet energy feature of the response signals obtained using the WPT technique. It was found that the energy-based WPT was successful not only in detecting the presence of the damage but also in efficiently recognizing the location of the damaged connectors. The energy rate of the analyzed signals by WPT was also considered by Asgarian et al. [318] as a damage indicator for an offshore platform. The successful performance of WPT in processing the vibration responses of a one-story RC frame was demonstrated by Chan et al. [319] under different seismic excitations. Nie et al. [320] used a hybrid damage identification approach based on a combination of WPT and principal component analysis (PCA) for a laboratory-scaled steel beam as illustrated in Figure 1.26. The accuracy of the proposed hybrid method in detecting the location of the damage was concluded.

1.4.3.5 Synchrosqueezed wavelet transforms

Synchrosqueezed wavelet transforms (SWT) are new WT-based techniques proposed by Daubechies et al. [321] that can efficiently analyze the nonlinear and nonstationary signal with high signal-to-ratio (SNR). The procedure of SWT is based on relocating the frequency-based information of the CWT coefficients to capture time-frequency components with higher resolutions.

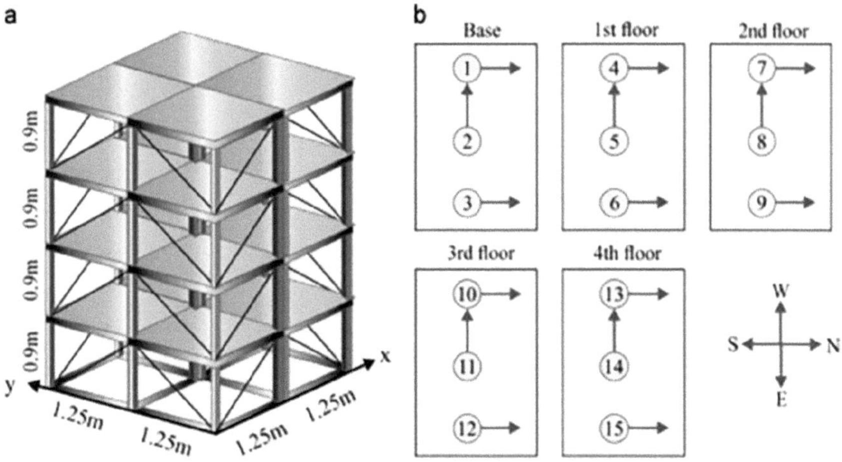

Figure 1.27 (a) Four-story test steel frame and (b) Location of the accelerometers on the structure [323].

In this method, first, the reconstruction of analyzed signals is performed by accumulating the features of signal components followed by determining the main frequencies of signals in time-frequency domains. Accordingly, the SWT can adequately detect the main frequencies of noisy signals. Wen et al. [322] proposed a combined application of SWT and SVM for fault diagnosis in a bearing system. In this method, first, the acceleration signals were decomposed into a series of IMFs using SWT. Then, the capability of the proposed hybrid SWT-SVM model in classifying the damage states of the bearing based on the energy feature of IMFs was concluded. A combination of SWT and the RDT was employed by Perez-Ramirez et al. [323] to identify the modal parameters of a 40-story steel frame as shown in Figure 1.27 including the natural frequencies and damping ratio.

A novel damage detection approach based on the combined actions of SWT and fractal dimension (FD) analysis named SWT-FD was proposed by Amezquita-Sanchez and Adeli [324] for damage detection and localization in a reduced-scale model of a 38-story concrete building structure under seismic excitations. The damage detection process of this model was based on the application of different algorithms of the FD technique on the measured vibration responses of the structure to identify the signal features followed by a denoising step of signals by SWT.

The SWT technique was introduced as an optimal signal processing technique by Wang et al. [325] in terms of consuming lower computational time in localizing the damage in a steel bar and a concrete structure. It was found that SWT can reduce the computational time without any considerable reduction in its performance accuracy. Li et al. [326] employed a combination of SWT and a multi-branch convolutional neural network

Figure 1.28 Experimental setup of field tests of a rail track [326].

on the AE waves of a rail track for detecting cracks in a railroad track and demonstrated its capability in detecting the crack existing not only on the surface of the track as shown in Figure 1.28, but also those generated inside the track. Su et al. [327] revealed the high efficiency of a new hybrid damage detection approach based on the use of SWT the application of SWT in combination with stack auto encoder algorithm for detecting the location of damage in composite plates when subjected to strong noise background.

1.4.3.6 Empirical wavelet transform

The EWT technique was introduced by Gilles [328] to solve the drawbacks of EMD such as low-frequency resolution leading to the inability of EMD in separating the signal components in which the ratio of low frequency of the signal to its high frequency is greater than 0.75. This advantage enables the EWT to efficiently analyze the nonlinear signals by reserving the original information of the analyzed signal and representing the results in time-frequency domains with high-resolution. The procedure of EWT is based on the decomposition of the original signal into a series of mode components by a suitable wavelet filter bank built using the Fourier spectrum. Due to the validated capability of the EWT in processing the nonlinear and nonstationary signals, the use of this technique has been extensively reported in the literature for damage detection in various structural and mechanical systems based on the analysis of the signal components captured through the application of the Fourier spectrum of signals. Despite the aforementioned advantages of the EWT in the analysis of nonlinear signals, it has a serious limitation in processing noisy signals that contain overlapping segments of the Fourier spectrum. To overcome this deficiency, several previous studies [329–332] have proposed different approaches to improve the performance of EWT by employing an optimal technique for segmentation of the Fourier spectrum components with district boundaries. Gilles and Heal [329] proposed a parameterless scale-space technique to simply and efficiently decompose the signal modes. Similarly, an adaptive parameterless EWT (APEWT) method was introduced by Zheng et al. [330] to improve the performance of the EWT in the proper segmentation of signals of a rotor system. The proposed method demonstrated superior fault diagnose performance compared to EMD-based methods. In a comparative study done by Liu et al. [331], EWT presented a more efficient approach to analyzing nonlinear signals of mechanical systems in comparison with the performance of EMD-based techniques. An efficient technique was used by Kedadouche et al. [332] for segmentation of the Fourier spectrum components of the EWT in the analysis of the signals of bearing systems. The superiority of the proposed method in fault diagnosis of the bearing system compared to the performance of the EMD and EEMD was concluded in terms of the accuracy of the model estimation and computational cost by adopting a DI based on the kurtosis feature of the IMFs decomposed by the EWT. A hybrid use of EWT and MUSIC techniques was employed by Amezquita-Sanchez [333] to present the time-frequency features of the analyzed signals with high-resolution. In the proposed method, the EWT with a wavelet filter bank was applied to the main signal frequencies determined by MUSIC in which the signal components had certain boundaries. The capability of EWT in capturing the fundamental frequency characteristics of distorted signals has been concluded by several previous studies [334,335]. A hybrid application of EWT in combination with HT, and STFT techniques was employed by Liu et al. [336] to obtain the time-frequency features of ultrasonic waves by performing a series of experimental tests. In the proposed method, the STFT is used for time-history representation of the signal modes identified by EWT followed by the application HT for the segmentation of the signal components. Despite the extensive application of EWT for fault diagnosis in mechanical systems, EWT has been rarely used for damage detection of civil engineering structures. A combination of EWT and the second-order blind identification (SOBI) method was proposed by Yuan et al. [337] to obtain the modal characteristics of laboratory-scale framed structures with tuned mass damper (TMD). In this method, the EWT was able to effectively identify the fundamental frequencies of structures with very close certain distances separated by SOBI. To identify the modal properties of a spatial frame structure such as natural frequencies, mode shapes, and damping ratios, Xin et al. [338] used an improved EWT technique. The superiority of the improved EWT method in characterizing the signal components was found in comparison with the performance of variational mode decomposition (VMD) technique. As another successful use of EWT combined with MUSIC and HT techniques was concluded by Amezquita-Sanchez et al. [339] in identifying the modal parameters of civil engineering structures such as natural frequencies and damping ratio. Mousavi et al. [340] demonstrated the successful performance of EWT in combination with the ANN algorithm for damage detection, localization, and quantification in a steel truss bridge model based on the evaluation of several signal features including energy, RMS, shape factor, kurtosis, and entropy extracted from the mode components by EWT.

1.4.4 S-Transform

The S-Transform (ST) technique is a combination of STFT and CWT techniques in which a Gaussian window with a frequency function [341,342]. Besides, this technique can represent an improved form of WT with a modification on the adopted mother wavelet. While performing this modification causes a deficiency associated with neglecting the wavelet's admissibility criterion leading to zero mean of the wavelet. Accordingly, this technique cannot be considered as the CWT algorithm. The procedure of ST is based on shifting down the window for certain frequency ranges varying in time. The majority use of ST has been previously reported for fault identification of simple systems rather than large-scale structures. Pakrashi and Ghosh [343] concluded the efficiency of the ST technique in analyzing noisy signals for fault diagnosis in a linear SDOF system by detecting any sudden stiffness changes, and crack localization in a steel beam with simple supports. The ST technique was employed by Ditommaso et al. [344] to analysis the blast-induced vibration response of a six-story RC building frame. The capability of the ST in damage localization in the structure was concluded through identifying the modal characteristics of the structure including the natural frequencies and mode shape parameters. Tehrani et al. [345] presented a successful use of the ST for damage identification and localization in a mass-spring system. An improved ST method with time-frequency representation was proposed by Liu et al. [346] for damage detection in a simply supported RC beam based on the analysis of the spikes that occurred in the energy feature. A ST-based signal processing technique based on the analysis of the peak responses of the modal parameters was introduced by Ponzo et al. [347] for damage identification in a five-story structure. Furthermore, Ponzo et al. [348] used S transform in damage detection and localization of RC frame subjected to an earthquake. A comparison study of using the S- Transform and Band-Variable Filter techniques are evaluated by Iacovino et al. [349] for damage localization based on the detection of curvature variations. Numerical data retrieved from nonlinear FE models subjected to damaging seismic excitation was used as a case study. Ditommaso and Ponzo [350] utilized the S-Transform in a damage assessment strategy of numerical and experimental RC frame models as shown in Figure 1.29.

Figure 1.29 Picture of RC frame model on the shaking table – University of Basilicata [350].

To improve the performance of the ST technique in terms of computational resource capacity required for the analyzed data, the fast S-Transform technique was proposed by Brown and Frayne [351]. This improvement is implemented by the fast ST technique by application of a down-sampling method in which a narrowed window is used to reduce the number of samples adopted. An application of the fast ST technique was reported by Ghahremani et al. [352] for damage identification in a six-DOF structure under Northridge earthquake excitation. It was found that the fast ST method can efficiently characterize the structural damage and its occurrence time through the analysis of the resonance frequencies of the structure. In addition, a combination of the fast ST and CNN algorithms was employed to classify the damage severity of the predefined damage states for the structure.

1.4.5 Wigner–Ville distribution (WVD)

The Wigner–Ville Distribution (WVD) technique is one of the time-frequency analyses which is known as a Cohen's class type representing high-resolution distributions in both time and frequency domains. Due to this capability of WVD, this technique is feasible for the analysis of nonlinear response signals of civil engineering structures. Although the primitive purpose of proposing this method was to solve the issue associated with statistical equilibrium in quantum mechanics [353], it has been extensively employed for SHM of structures. Despite the advantages of WVD such as simplicity, high efficiency in the analysis of nonlinear signals, time-frequency distributions with

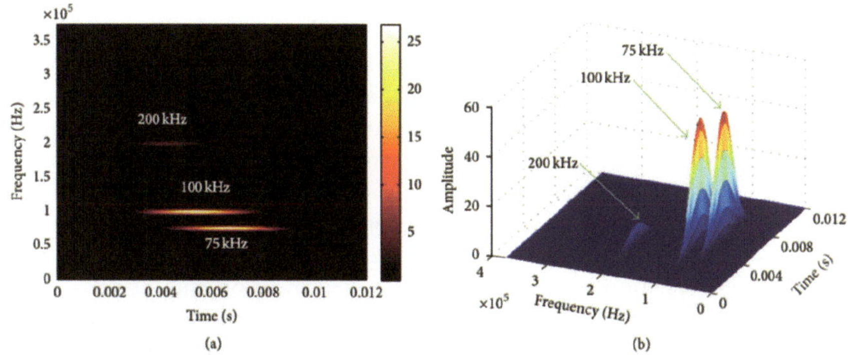

Figure 1.30 The SPWVD representation of the sample signal: (a) two-dimensional figure and (b) three-dimensional figure [362].

high-resolution, needless to adopt any window function, etc. it may generate fictitious and spurious frequency due to its deficiency in representing any appropriate cross-term interference. The application of WVD in the monitoring of mechanical systems has been extensively reported in the literature [354–357]. Besides, the WVD has been also used for extracting the natural frequencies of signals [358], and identifying the structural damage location using the Non-Destructive Testing (NDT) technique by adopting ultrasonic guided waves [359]. A damage detection approach based on the use of WVD was proposed by Katunin [360] to be applied for FE models of aluminum beams. The proposed method was properly able to detect the presence and quantify the severity of the damage in the beams. Qatu et al. [361] employed the WVD technique for denoising purposes and presented high-resolution distributions of the signal energy in time-frequency domains. It was found that this technique not only could identify the existence of the damage in an aluminum plate but also was capable of efficiently detecting the damage location. A hybrid use of an improved WVD-based algorithm named Smoothed Pseudo WVD (SPWVD) and a peak-track algorithm was proposed by Wu et al. [362] for damage identification in a carbon fiber composite material. As a result, as shown in Figure 1.30, the proposed approach could accurately represent the time-frequency distributions of the analyzed signal obtained from the damaged cases with high-resolution.

1.5 DYNAMIC FEATURE EXTRACTION METHODS BASED ON MACHINE LEARNING

In recent years, the SHM applications of machine learning (ML) as a subset of artificial intelligence (AI) have increased in combination with various signal processing techniques for feature extraction of response data of civil engineering structures. ML techniques are categorized into (i) supervised learning techniques which include classifiers such as Bayesian method, deep learning (DL), SVM, etc., and regression-based methods such as decision trees, and neural networks, (ii) unsupervised learning or clustering methods such a k-means, spectral clustering, partitional clustering, Hierarchical clustering, etc., and (iii) reinforcement learning techniques in which the performance of the model is optimized by performing try-and-error technique. Generally, the procedure of ML techniques in vibration-based SHM applications such as SVM and ANN techniques is based on feature extraction and data classification resulting in damage identification and localization.

DL techniques are also known as practical ML techniques for SHM of complex and large-scale structures with big and large datasets, which can improve the performance of ML by eliminating the pre-selection procedure of suitable features. These techniques also basically work to automate the feature extraction procedure through deep neural network (DNN) layers each of which deals with learning a new representation of the input response data. Due to simultaneous performances of the feature extraction and data classification procedures in DL methods, these techniques are more practical in SHM since minimum knowledge is required about the specific features as shown in Figure 1.31. Compared to traditions data-driven SHM techniques which require proficient knowledge about the features in addition to a step-by-step training procedure, DL-based SHM presents an automatic feature extraction procedure as shown in Figure 1.31 in the form of an end-to-end design simultaneously performed with training procedure through DNN layers which are efficiently compatible with the analysis of large datasets.

There exist several traditional methods and modern techniques in the field of dynamic feature extraction in time and frequency domains based on ML and DL. Since the CNN algorithm adopts the trained convolutional layers without any maximum pooling layer, noise filtering, and retaining

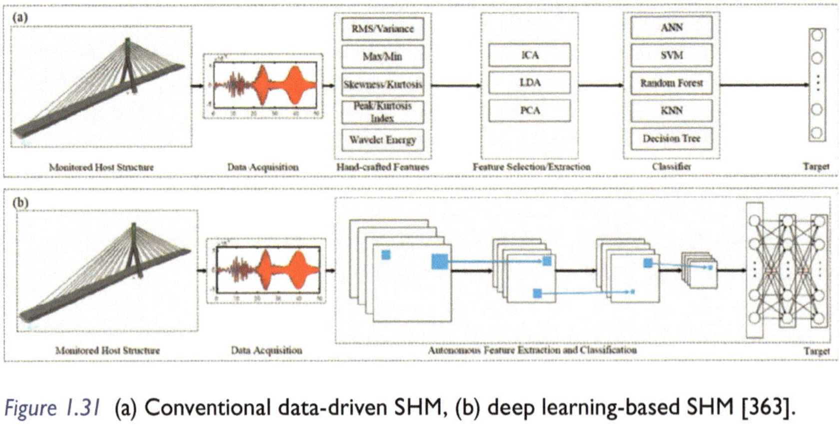

Figure 1.31 (a) Conventional data-driven SHM, (b) deep learning-based SHM [363].

the predominant signal frequency, this technique can accurately analyze and train noisy input data. Hence, this technique has been widely used for estimating the responses of different structural systems. An FE-based convolutional neural network (CNN) algorithm was trained by Lin et al. [363] as a DL technique to analyze the response data of a simply supported beam using a series of signal features including the signal modes and frequency bands. A two-dimensional (2D) CNN method was adopted by Wang and Cha [364] for feature extraction of acceleration response data normalized by CWT and FFT techniques. In addition, the One-Class Support Vector Machines (OC-SVM) were utilized for localizing the fault and the loosen bolts as a novel detector. The CNN technique was also employed by Wu and Jahanshahi [365] for estimating the dynamic responses of SDOF MDOF systems and its superiority was concluded in comparison with the Multilayer Perceptron (MLP) technique in analyzing and training of noisy data. As an improved neural network technique compared to ANNs, a two-step procedure was proposed by Pathirage et al. [366] based on the combination of dimension reduction and relationship learning DNNs. The proposed procedure began with the feature extraction of the inputted data by using a deep autoencoder, then, they were defined to the network for damage identification. To reduce the effects of noise, a novel frequency-domain convolutional neural network (FDCNN) based on the Bouc–Wen hysteretic model and utilizing a spectral pooling operator, was proposed by Lopez-Pacheco et al. [367] for structural fault diagnosis. It was concluded that the proposed technique could significantly reduce the computation time compared to time-domain networks.

Besides, time-domain networks have been also widely employed by previous studies in the literature. Pan et al. [302368] and Pan and Yang [369] proposed the application of the CNN algorithm based on the parallel use of the one-dimensional (1D) and WPT methods for the extraction of time-domain and time-frequency-domain features, respectively. As an improvement on the proposed framework by Bao et al. [370] based on the use of deep convolution neural networks (DCNNs), Tang et al. [371] proposed the use of a parallel procedure for extracting the time- and frequency-domain features of balanced response data. Accurate performance of the 2D CNN technique was concluded by Khodabandehlou et al. [372] for damage classification of a highway bridge from the analysis of the acceleration responses of the bridge under shake table vibrations. From a comparative study on the performances of the CNN technique, k-nearest neighbor (KNN), SVM, decision tree (DT), and random forest methods in damage identification of a cable bridge by Li and Sun [373], more accuracy of the CNN was obtained compared to other methods. Li et al. [374] employed a 1D-CNN method for damage classification of a bridge model by defining deflection responses as input data. The superiority and more accuracy of CNN in estimating the structural responses were concluded in comparison with the traditional methods such as SVM, KNN, and DT. Besides, Hung et al. [375] evaluated the performance of a hybrid technique from the combination of 1D-CNN and Long-Short Term Memory (LSTM) network for damage identification using TS data.

As discussed above, since the CNN technique adopts a noise filtering algorithm to increase the SNR of noisy raw data and facilitate the subsequent peak detection, this technique has been widely used for vibration-based SHM of structures in noisy environments with low SNRs. There exist further studies in the literature proposing solutions for measurements under low SNR based on ML and DL techniques [376,377]. Ding et al. [376] demonstrated the successful application of a sparse Deep Belief Network (DBN) which is based on Restricted Boltzmann Machines (RBM) in detecting and locating the damage by utilizing a powerful denoising algorithm. Also, the superiority of the proposed technique was concluded compared to the performance of swarm intelligence techniques. A modified version of Residual Convolutional Neural Network (ResNet) was proposed by Fan et al. [377] to denoise the acceleration response data of a TV tower before identifying the modal responses using dropout, skip connection, and sub-pixel shuffling layers. Besides, a hybrid framework from the combination of ML and wavelet transform technique was used by Ying et al. [378] for damage identification in a steel pipe under environmental vibrations based on ultrasonic measurements. The proposed framework was able to accurately estimate the presence of the damage by denoising the raw data before wavelet processing. Shang et al. [379] proposed a damage detection approach based on an unsupervised DNN in which a deep convolutional denoising autoencoder was utilized to obtain effective signal features that are sensitive to the damage in a simply supported beam under the effects of noise and temperature. The proposed method efficiently estimated the modal responses of the beam by preserving the original information of the response data inputted as the cross-section functions. A hybrid damage detection technique was also used by Rafiei and Adeli [380] for SHM of a high-rise building structure from the combination of wavelet transform and FFT as the signal processing techniques, with unsupervised ML algorithms including RBM and neural dynamics classification (NDC) algorithms. The high-accuracy performance of the proposed approach in damage identification and classification was concluded in comparison with the performances of several conventional supervised classification algorithms when utilizing denoized responses as the input of NDC.

1.6 ADVANCED SIGNAL PROCESSING TECHNIQUES FOR SHM

1.6.1 Background of spectral analysis approaches

SHM of strategic civil engineering structures such as bridges using signal processing techniques has been one of the most important concerns of engineers and researchers over the past few decades. Very extensive efforts have been made in the literature in exploring and improving the signal processing techniques for damage identification of structural and mechanical systems. Due to existing limitations in the processing algorithm of traditional signal processing techniques with frequency representation such as FFT [97,98,381–383], FDD [164,165], and FRF [111–115], these approaches have encountered serious difficulties in the process of nonlinear and nonstationary signals from real-word complex structures in noisy environments.

Although the use of signal processing techniques with time-frequency representation such as STFT [174,212], Wigner–Ville distribution (WVD) [354], WT [229–231], CWT [341,342], Hilbert–Huang Transform (HHT) [232–235] have significantly improved the damage detection process of nonlinear and nonstationary signals through enhancing the accuracy of feature extraction, they still have some unsolved difficulties as mentioned follows.

Most of the primary signal processing approaches are based on the fact that the signal is linear and stationary. However, in general, the signals are often nonlinear and nonstationary in the real world. Since the time-frequency techniques can supply both time and frequency details of a signal, they are appropriate to process nonlinear and nonstationary signals. In spite of the higher extraction efficiency of the classical FFT compared to other algorithms, it has been shown that it is deficient in the processing of nonlinear and nonstationary signals. Also, STFT is basically the windowed Fourier transforms which can be used to analyze nonstationary signals and linear data. However, these methods lack multiresolution analysis capabilities.

Due to some key advantages of frequency-domain methods based on the Fourier transformation such as reducing the data volume and repairing the lost information by averaging the effects of random noise, these methods have been widely utilized in the literature. One of the most popular frequency domain methods is a method based on the FRF in which a ratio between the frequency domain response of the structure to the applied force is generated to estimate the modal parameters of the structure using curve fitting methods [384]. Maia et al. [385] employed FRF in damage detection of a simplified system and also a real bridge structure [116]. More efficiency of the FRF method in detecting the damage during the first resonance of the bridge response based on the features including the shift in the resonance and the amplitude changes was concluded compared to the performance of the conventional modal parameters such as natural frequencies, modal shapes, modal damping. The capability of the FRF in damage detection of various structures has been previously obtained in experimental [133] and numerical [386387] studies. Liu et al. [122] revealed that although FRF works only for low-frequency range, the use of FRF shapes caused a more clear damage localization approach in a cantilever beam compared to conventional mode shapes. In addition, the performance of FRF in damage detection of framed structures was evaluated by Mohan et al. [128]. It was concluded that the FRF method had better performance in comparison with GA, and particle swarm optimization (PSO) methods due to presenting mode shape information in addition to natural frequencies by FRF. An FRF-based interpolation method was proposed by Dilena et al. [388] to detect and locate the damage in a bridge structure compared to the modal curvature method. Lee and Kim [389] proposed damage indices based on the changes generated in the FRF shapes and SFRF. They concluded the superiority more explicitly of the FRF method applied to dynamic data compared to the flexibility-based approach extracted from the modal data. Besides, the output-only methods in the frequency domain named frequency domain decomposition techniques (FDD) were developed by Brincker et al. [158,390] to improve the existing drawbacks of BFD or PP method in identifying the modal parameters of closely spaced modes. The MUSIC algorithm provides a high-resolution power density spectrum which effectively improves the detectability of close frequencies in comparison with the performances of traditional techniques such as FFT in analyzing signals with low SNRs [173]. Therefore, as an extension of the previous research works, a hybrid damage detection approach is proposed in this study from the combined use of the MUSIC algorithm with CEEMDAN for damage detection of a steel truss bridge. Jiang and Adeli [177] proposed a new approach based on the MUSIC algorithm for a scaled model of a 38-story RC structure. The results of the output predicted by the trained neural network model and the output measured by the sensors were compared and the difference between them was considered as an effective indicator of damage. Rios and Sanchez [178] used a fusion of MUSIC-ANN algorithms to address the location of damage and quantify the damage severity of a five-bay truss structure. To do that, the amplitude and the natural frequencies of the structural responses were used as the input data in an ANN. The result showed that the proposed method was effectively simple to use, automated, and reliable tool in SHM. MUSIC algorithm has also been used in mechanical structures. For instance, Perez et al. [182] proposed a fault detection procedure based on MUSIC for the purpose of high-resolution spectral analysis to enhance the anomalies detectability in an induction motor. Martinez et al. [183] combined EMD and MUSIC techniques to detect the mixed faults in induction motors. The results showed the advantage of the proposed approach which could identify the characteristic frequencies of the multiple combined faults.

Although the classical Fast Fourier spectral analysis and frequency-based methods have a higher extraction efficiency than other algorithms, it has been shown that it is deficient in the processing of nonlinear and nonstationary signals. The traditional time-frequency analysis methods, such as STFT are basically the windowed Fourier transforms which can be used to analyze nonstationary signals and linear data. However, these methods lack capabilities in multiresolution analyses. Unlike the short-time Fourier transform, the Wavelet transform is a powerful tool that includes the incorporating of variable window sizes, and capabilities in resolution, thus balancing both time and frequency domains of a signal. Similar to STFT, WT can only be used to process signals in nonstationary and linear conditions.

To overcome the aforementioned drawbacks of the above-mentioned methods, the Hilbert–Huang transform was developed by Huang et al. [232,390], which is promising in the processing of nonlinear and nonstationary signals. This approach includes two procedures: In the first part, the signal is decomposed by the EMD method into a finite set of intrinsic oscillatory components, called intrinsic mode functions (IMFs), which reflect the scale characteristics embedded in the signal. In the second part, the HT is used to obtain the instantaneous frequency and amplitude features and yield a time-frequency representation (Hilbert spectrum) for each IMF. So far, this method has been applied in different fields of engineering, such as earthquake engineering, SHM, and damage detection. Vincent et al. [236] utilized EMD and WT methods as signal processing techniques in damage detection and revealed that EMD is a more efficient method than WT, due to its novel algorithm. Quek et al. [391] utilized the Hilbert–Huang Transform to assess the nonstationary wave propagation fundamentals of different structural members, including aluminum beams, sandwiched aluminum beams, plates, and RC slabs, by considering various damage levels. Yang et al. [392] employed the EMD and HT to determine the damage time instant, the natural frequencies, damping ratios, and damage locations, before and after damage, in which damage scenarios were accomplished by applying sudden changes in the stiffness of an ASCE benchmark structure. In that study, the presence of damage was recognized by observing the damage spikes in the signals using EMD. Xu and Chen [243] conducted an experimental investigation to detect and locate damage using the EMD method by applying abrupt changes in the structural stiffness of a three-story shear building. Pines and Salvino [247] proposed a signal processing method based on the processing of TS data from a 1-D scaled civil building, tested in the cases with and without structural damage. From the results, it was found that this method is capable to address the unique features of the vibratory response of the structure. Liu et al. [393] demonstrated that the Hilbert–Huang Transform method can efficiently detect and locate the possible damage in a scaled four-story structure by comparing the frequency of the first three IMFs extracted before and after damage states.

Yan and Miyamoto [394] evaluated the modal frequencies and damping ratios on a Z24-bridge by comparing HHT and continuous wavelet transform (CWT) methods. Li et al. [395] utilized the combination of EMD and WT methods to predict the location and intensity of the damage. At first, the acquired signal from the vibration of a four-story shear structure was decomposed into a set of IMFs using the EMD method, then, the WT coefficients detected the location and severity of the damage. Cheraghi and Taheri [396] proposed some numerical and experimental monitoring approaches to identify the various damage levels, detect and locate damage in the pipelines by evaluating the EMD-based energy damage indices for the vibration response captured using piezoelectric sensors. Also, Rezaei and Taheri [397] employed the EMD on the experimental vibration responses of steel pipes to detect damage by defining an energy-based DI. It was found that the EMD energy-based DI is a sensitive and effective method for detecting and locating damage. Chen [262] summarized the application of the HHT technique as a powerful tool in SHM. The efficiency of this approach in data processing, identification of structural parameters, and damage detection of tall buildings were evaluated by conducting some experimental and numerical studies.

Bao et al. [181] applied an improved HHT algorithm as a signal processing tool to identify the damage conditions of a scaled concrete-steel composite beam model subjected to impact tests. Tang et al. [105] proposed a damage detection index, named as the ratio of equivalent damping (RED), which was evaluated using HHT and FFT methods, to assess the shaking table test data obtained from the benchmark models. Subsequently, Hsu et al. [398] evaluated the sensitivity of the DI proposed by Tang et al. [105] for analyzing the initial damage in cantilever beams and a full-scale 3D three-story steel frame structure subjected to earthquakes. Shi et al. [257] used the combination of two efficient techniques including the RDT in the time-domain, and the HHT method in the time-frequency domain, to identify the modal frequencies and damping ratios of the Shanghai World Financial Center (SWFC) in China. Garcia-Perez et al. [258] presented a fusion methodology of WPT and EMD, combined with ANN, to identify and locate the combined damage in a scaled model of a five-bay truss-type structure. Meredith [264] studied the possibility of applying EMD, and used a moving average filter on the acceleration response of a beam subject to a moving load, to detect the damage location. To verify this technique, they conducted several scenarios including a range of bridge lengths, speeds of the moving load, and noise levels. Carbajo et al. [399] presented a new automated structural change detection algorithm by extraction of DSFs of instantaneous frequencies and signal energies from HHT, that were applied in an experimental single degree of freedom (DOF) system and a wind turbine blade under band-limited base excitation. Dushyanth [400] evaluated the HHT method to pinpoint the location of damage, in both simulation and experimental tests, and used a set of damage indices and a multi-level SVM, to distinguish the healthy or damaged state of an aluminum plate. The authors concluded that the proposed detection method can be applied to real-time applications due to its high accuracy and the considerable decrease in computational time. Xun and Yan [401] used a radial basis function (RBF) neural network as a pre-processor to expand the length of the signal for removing the end swings problem and as a post-processor to select the optimal IMFs.

1.6.2 Approaches to overcome some limitations of EMD-based techniques

In spite of the major applications of EMD as a powerful signal processing tool in the literature, it has a significant inherent defect called mode mixing due to the intermittent frequencies in the IMF. To overcome this, Wu and Huang [271] proposed the EEMD which was an improved method through the addition of identically white Gaussian noise, with an appropriate scale and standard deviation. Martinez [184] combined the EEMD with the MUSIC approach and compared it with traditional methods such as DWT and FFT techniques as a signal processing tool to identify the modal frequencies of structures, and validated the approach by experimental results from a scaled truss bridge. The results demonstrated the efficiency of the proposed methodology. Amiri and Darvishan [276] evaluated the capability of EEMD compared with EMD using a density-based clustering technique. This comparison was implemented by utilizing the frequency and amplitude features extracted from the numerical acceleration response of a steel moment frame.

Unfortunately, the EEMD has two major drawbacks: residual noise in IMF, and difficulty in averaging of a probably different number of IMFs caused by adding different Gaussian white noise to the signal. In 2011, Torres et al. [273] proposed the CEEMDAN to solve the EEMDs' faults. The CEEMDAN procedure has a different approach from EEMD, by adding a particular noise, at each stage, instead of adding white Gaussian noise, which is obtained with a unique residue after each IMF extraction. Additionally, it has a proper spectral separation, increasing the level, and reducing the time of decomposition. Recently, Mousavi et al. [224] proposed a combination of CEEMDAN-Hilbert transform with artificial neural networks named CEEMDAN-HT-ANN to identify the damage and classify its severity in a laboratory-scale model of a steel truss bridge exposed to white noise excitations using different damage indices based on the extracted features from the IMFs including energy, IA, unwrapped phase, and instantaneous frequency (*IF*). The capability of the proposed damage detection approach was successfully concluded and the energy-based DI performed more efficiently in detecting the damage compared to those based on the other features. The three aforementioned EMD-based techniques have been widely used in mechanical [277–281], geological, seismic signals [282], and electrocardiogram (ECG) [273] applications.

EMD technique is able to decompose any vibration signal into a set of simple IMF. Each of IMF has two conditions. The first one considers the number of extrema and zero-crossing either equal or differs only by one. In the second condition, the mean of the envelope of local minimum and the local maximum is zero (symmetrically) and it is locally symmetric. The theoretical basics of the EMD can be found in Huang et al. [232]. Despite the major applications of EMD, it has a significant defect called mode mixing. To overcome this, Wu and Huang [271] proposed the EEMD which was an improved method by adding an identically white Gaussian noise with appropriate scale and standard deviation. EEMD can diminish the mode mixing problem, by adding Gaussian white noise to the input signal. However, there exist some shortcomings, such as independence of the decomposition process of a sampled signal. This can be caused by the residual noise in IMF and deficiencies in the decomposition procedure. Also, there is a difficulty in averaging of a different number of IMFs caused by adding different white Gaussian noise to the signal. To overcome these drawbacks, CEEMDAN was proposed as described in the following Section 2.3.

1.6.3 Empirical mode decomposition

EMD is able to decompose any vibration signal to a combination set of simple intrinsic mode functions referred to 'IMF'. Each IMF has two conditions. The first one considers the number of extrema and zero-crossing either equal or differs only by one. In the second condition, the mean of the envelope of local minimum and the local maximum is zero (symmetrically) and it is locally symmetric. Decomposition of the original signal into IMFs is done via 'sifting process' and can be defined as follows:

1. Finding all extrema (maxima and minima) of the TS of the signal, $x(t)$.
2. Generating a cubic spline on the upper and lower envelope by connecting the maximum and minimum points [$e\max(t)$, $e\min(t)$].
3. Computing the difference of local mean $d(t)=x(t)-r(t)$; where $r(t)=[e\min(t)+e\max(t)]/2$.
4. Iterating the computation of $r(t)$ and $d(t)$ to the end of the decomposition process which leads to final $x(t)$ from the sum of IMFs plus the final residual as follows:

$$x(t)=\sum_{n=1}^{N} d_n(t)+r_N(t) \tag{1.1}$$

where $d_n(t)$ is the *nth* IMF, and $r_N(t)$ is the final residual.

1.6.4 Ensemble empirical mode decomposition

Despite numerous applications of EMD, it has a big defect called mode mixing. To overcome this, Wu and Huang [271] proposed the EEMD as an improved method by adding an identically white Gaussian noise with an appropriate scale and standard deviation. EEMD algorithm can be described by the following steps:

1. Add the white Gaussian noise, $w_i(t)$ ($i=1, 2, \ldots, I$), to the initial signal, $x(t)$:

$$x_i(t)=x(t)+\beta w_i(t)$$

where β indicates the amplitude of the ith added white noise, and I is the ensemble size.

2. Decompose the signal $x_i(t)$ to obtain the IMF, $c_{ji}(t)$ where $j=1, 2, \ldots, N$, and $c_{ji}(t)$ is the jth IMF of $x_i(t)$.
3. Compute the ensemble mean of the corresponding component as a final decomposed IMFs:

$$\bar{c}_j(t) = \frac{1}{I}\sum_{i=1}^{I} c_{ji}(t) \tag{1.2}$$

where $\bar{c}_j(t)$ is a mean of jth IMF of $x(t)$.

EEMD can diminish the mode mixing problem, by adding the white Gaussian noise to the input signal. However, there exist some shortcomings such as independence of the decomposition process of $x_i(t)$ for I times. This can be caused by the residual noise in IMF and deficiencies in the decomposition procedure. Also, difficulty in averaging of a probably different number of IMFs caused by adding different white Gaussian noise to the signal. To overcome these drawbacks CEEMDAN was proposed by Torres et al. [273] as follows.

1.6.5 Complete Ensemble Empirical Mode Decomposition with adaptive noise

The EEMD has two major drawbacks: (i) residual noise in IMF and (ii) difficulty in averaging of a probably different number of IMFs caused by adding different white Gaussian noise to the signal. Torres et al. [273] solved these faults by proposing the CEEMDAN. The CEEMDAN has a different trend from EEMD in adding a particular noise $E_j(w_i(t))$ at each step of the decomposition process instead of adding the white Gaussian noise which is obtained with a unique residue after each IMF's' extraction.

The CEEMDAN method can be defined as the following steps:

1. Add $E_1(w_i(t))$ to the initial signal, $x(t)$, where w_i indicates the ith added white Gaussian noise.

$$x_i(t) = x(t) + \beta_0 E_1(w_i(t)) \tag{1.3}$$

2. Calculate the first $IMF(\bar{c}_1(t))$ through the first residue $(r_1(t))$ as follows:

$$\bar{c}_1(t) = x(t) - r_1(t), \text{ where } r_1(t) = \frac{1}{I}\sum_{i=1}^{I} M(x_i(t)) \tag{1.4}$$

where $M(\cdot)$ is the operator that represents the local means of a signal.

3. Obtain the second IMF $\overline{c}_2(t) = r_1(t) - r_2(t)$, where $r_2(t) = \frac{1}{I}\sum_{i=1}^{I} M\left(r_1(t) + \beta_1 E_2(w_i(t))\right)$, and $E_2(w_i(t))$ is the second IMF of EEMD and I is ensemble size.
4. Repeat step 3 to obtain jth IMF of CEEMDAN $\overline{c}_j(t) = r_{j-1}(t) - r_j(t)$ where

$$r_j(t) = \frac{1}{I}\sum_{i=1}^{I} M\left(r_{j-1}(t) + \beta_{j-1} E_j(w_i(t))\right) \tag{1.5}$$

where $\beta_j = \varepsilon_0 \text{std}(r_j(t))$ is the SNR. Due to the empirical nature of the EMD, implementing a series of trial-and-error tests is needed to achieve a proper decomposition of the signal. For this reason, CEEMDAN, EEMD, and EMD methods converge based on noise standard deviation (Nstd), number of realizations (NR), and the maximum number of sifting iterations (MaxIter). The converge values for these parameters are Nstd= 0.2, NR= 100, and MaxIter= 1000.

1.6.6 The Hilbert–Huang transform (HHT)

The Hilbert–Huang transform (HHT) developed by Huang et al. [234] as an adaptive and capable tool in the nonstationary and nonlinear situations includes two procedures. In the first procedure, EMD decomposes a signal into a finite set of intrinsic components called "intrinsic mode function" (IMF). Then, the HT is applied to each IMF to obtain instantaneous frequencies (IF) and IA features. This eventually yields a time-frequency representation (Hilbert spectrum) for each IMF. According to [234], for a signal, $X(t)$, HT is obtained, $Y(t)$, as:

$$Y(t) = \frac{1}{\pi} P \int_{-\infty}^{\infty} \frac{X(t')}{t - t'}\, dt' \tag{1.6}$$

where P is the Cauchy principal value, and t is the time variable and t' is a time interval.

Then the HHT of the signal can be given by

$$Z(t) = X(t) + iY(t) = a(t)e^{i\theta(t)} \tag{1.7}$$

where $a(t)$ and $\theta(t)$ are the amplitude and phase, respectively, which are defined as:

$$a(t) = \sqrt{X^2(t) + Y^2(t)},\ \theta(t) = \arctan\left(\frac{Y(t)}{X(t)}\right) \tag{1.8}$$

Also, the instantaneous frequency is the derivative of the phase function:

$$f(t) = \frac{1}{2\pi}\frac{d\theta}{dt} \tag{1.9}$$

Equations (1.8) and (1.9) are used to compute the instantaneous amplitudes and frequencies of each IMF as an input parameter in the EMD-based damage index.

1.6.7 Empirical wavelet transform

The EWT was proposed by Gilles [328] as a novel adaptive decomposition approach based on the combination of EMD and WT techniques to more accurately present the local characteristics of the signals than Fourier transform (FT). The theory of the EWT algorithm is based on the use of the Mayer wavelet during the reconstruction process of the analyzed signal. In this method, the Fourier spectrum is segmented into N-number contiguous segments (ω_N), $\Lambda_n = [\omega_{n-1}, \omega_n]$ represents each segment between the limits of $\omega_0 = 0$, and $\omega_N = \pi \left(\text{i.e., } \bigcup_{n=1}^{N} \Lambda_n = [0, \pi] \right)$. Around the center of each segment, a transition phase T_n with a width of $2\tau_n$ is defined and the empirical wavelet is employed on each segment Λ_n as the band-pass filter as shown in Figure 1.32 When $\forall n > 0$, the empirical scaling function and the empirical wavelets are defined as expressed in Equations (1.1) and (1.2), respectively.

$$\hat{\phi}_n(\omega) = \begin{cases} 1 & \text{if } |\omega| \leq (1-\gamma)\omega_n \\ \cos\left[\frac{\pi}{2}\beta\left(\frac{1}{2\gamma\omega_n}\left(|\omega|-(1-\gamma)\omega_n\right)\right)\right] & \text{if } (1-\gamma)\omega_n \leq |\omega| \leq (1+\gamma)\omega_n \\ 0 & \text{otherwise} \end{cases} \tag{1.10}$$

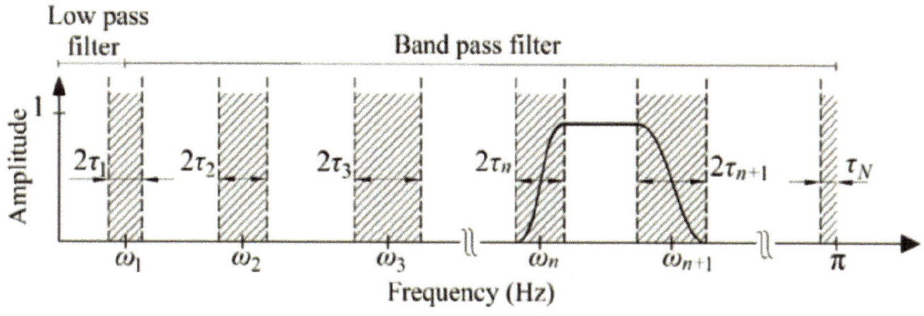

Figure 1.32 Contiguous segments of the Fourier spectrum developed by the EWT algorithm.

$$\hat{\psi}_n(\omega)=\begin{cases}1 & \text{if } \omega_n+\tau_n\leq|\omega| \leq\omega_{n+1}-\tau_{n+1}\\ \cos\left[\frac{\pi}{2}\beta\left(\frac{1}{2\tau_{n+1}}\left(|\omega|-\omega_{n+1}+\tau_{n+1}\right)\right)\right] & \text{if } \omega_{n+1}-\tau_{n+1}\leq|\omega| \leq\omega_{n+1}+\tau_{n+1}\\ \sin\left[\frac{\pi}{2}\beta\left(\frac{1}{2\tau_n}\left(|\omega|-\omega_n+\tau_n\right)\right)\right] & \text{if } \omega_n-\tau_n\leq|\omega| \leq\omega_n+\tau_n\\ 0 & \text{otherwise}\end{cases} \tag{1.11}$$

where the function β (x) is an arbitrary function C^k ([0,1]) given as follows:

$$\beta(x)=\begin{cases}0 & \text{if } x\leq 0\\ 1-\beta(1-x) & \forall x\in[0,1]\\ 1 & \text{if } x\geq 1\end{cases} \tag{1.12}$$

As recommended in [328], the simplest option to obtain τ_n is choosing the proportional relation between τ_n and ω_n such that $\tau_n=\gamma\omega_n$ where $0<\gamma<1$. Therefore, considering $\forall n>0$, Equations (1.11) and (1.12) can be re-written as Equations (1.4) and (1.5), respectively.

$$\hat{\phi}_n(\omega)=\begin{cases}1 & \text{if } |\omega|\leq (1-\gamma)\omega_n\\ \cos\left[\frac{\pi}{2}\beta\left(\frac{1}{2\gamma\omega_n}\left(|\omega|-(1-\gamma)\omega_n\right)\right)\right] & \text{if } (1-\gamma)\omega_n\leq|\omega| \leq(1+\gamma)\omega_n\\ 0 & \text{otherwise}\end{cases} \tag{1.13}$$

$$\hat{\psi}_n(\omega)=\begin{cases}1 & \text{if } (1+\gamma)\omega_n\leq|\omega|\leq(1-\gamma)\omega_{n+1}\\ \cos\left[\frac{\pi}{2}\beta\left(\frac{1}{2\gamma\omega_{n+1}}\left(|\omega|-(1-\gamma)\omega_{n+1}\right)\right)\right] & \text{if } (1-\gamma)\omega_{n+1}\leq|\omega|\leq(1+\gamma)\omega_{n+1}\\ \sin\left[\frac{\pi}{2}\beta\left(\frac{1}{2\gamma\omega_n}\left(|\omega|-(1-\gamma)\omega_n\right)\right)\right] & \text{if } (1-\gamma)\omega_n\leq|\omega|\leq(1+\gamma)\omega_n\\ 0 & \text{otherwise}\end{cases} \tag{1.14}$$

Based on the theory of building a frame set of empirical wavelets described above, the EWT, $W_f^{\varepsilon}(n,t)$, is defined in the same manner as for the classic wavelet transform. The detail coefficients are given by the inner product with the empirical wavelets:

$$W_f^{\varepsilon}(n,t)=\langle f, \psi_n \rangle=\int f(\tau)\,\overline{\psi_n(\tau-t)}\,d\tau=\left(\hat{f}(\omega)\overline{\hat{\psi}_n(\omega)}\right)^{\vee} \tag{1.15}$$

And the approximation coefficient, $W_f^{\varepsilon}(0,t)$, is defined by the inner product with the scaling function given as follows:

$$W_f^{\varepsilon}(0,t)=\langle f, \phi_1 \rangle=\int f(\tau)\,\overline{\phi_1(\tau-t)}\,d\tau=\left(\hat{f}(\omega)\overline{\hat{\phi}_1(\omega)}\right)^{\vee} \tag{1.16}$$

where $\hat{\psi}_n(\omega)$ and $\hat{\phi}_n(\omega)$ can be obtained using Equations (1.13) and (1.14), respectively.

Therefore, the reconstruction can be obtained as follows:

$$\begin{aligned} f(t) &= f_0(t) + \sum_{n=1}^{N} f_n(t) = W_f^{\varepsilon}(0,t) * \phi_1(t) + \sum_{n=1}^{N} W_f^{\varepsilon}(n,t) * \psi_n(t) \\ &= \left(\widehat{W_f^{\varepsilon}}(0,\omega)\hat{\phi}_1(\omega) + \sum_{n=1}^{N} W_f^{\varepsilon}(n,\omega) * \hat{\psi}_n(\omega)\right)^{\vee} \end{aligned} \tag{1.17}$$

where * denotes the convolution operators.

1.6.8 FDD algorithm

FDD technique is one of the popular frequency domain algorithms of OMA which was proposed by Brincker et al. [94] to overcome the existing drawbacks of BFD or PP method in identifying the modal parameters of closely spaced modes. In this method, the modal parameters of the output-only systems are estimated using the SVD based on the matrix of output PSD. With assuming that a white noise is an input in OMA algorithms and the PSD is constant, the relationship between the inputs and the output in the FDD method can be written as follows:

$$\left[G_{yy}(\omega)\right] \propto \left[H(\omega)\right]\left[I\right]\left[H(\omega)\right]^{H} \tag{1.18}$$

where $[G_{yy}\,(\omega)]$ is the output PSD matrix for the frequency of ω, $[H\,(\omega)]$ is the FRF matrix and the superscript H denotes the Hermitian of a matrix, and $[I]$ an identity matrix. Since the PSD matrix of outputs is decomposed in FDD, the SVD of $[G_{yy}(\omega)]$ for a specific frequency of ω_k can be written as follows:

$$\left[G_{yy}(\omega_k)\right] \propto [U][S][V]^H \tag{1.19}$$

where $[U]$ and $[V]$ are unitary matrices holding the singular vectors, and $[S]$ is the matrix of singular values arranged in descending order.

Since the PSD matrix is a square Hermitian matrix and the singular vector matrices of $[U]$ and $[V]$ are theoretically the same, the SVD of $[G_{yy}(\omega)]$ can be re-written as follows:

$$\left[G_{yy}(\omega_k)\right] \propto [U][S][U]^H \tag{1.20}$$

In the FDD method, the modal frequency is obtained using a PP approach in which the singular vectors of $[U]$ are employed to estimate the mode shapes. That is, for a given frequency, the number of non-zero singular values denotes the number of predominant modes on the responses of the system.

1.6.9 MUSIC algorithm

The MUSIC algorithm can detect frequencies with high-level noise with a high-resolution spectrum method. In this method, first, a discrete-time signal $x(t)$ is considered as the sum of p complex sinusoid and noise $x_n(t)$ given as follows:

$$x(t) = \sum_{k=1}^{P} I_k e^{j(2\pi f_k + \phi_k)} + x_n(t) \quad n = 0,1,2,\ldots,N-1 \tag{1.21}$$

where N, I_k, f_k, and φ_k are the number of sample data, amplitude, frequency, and the phase of the k-th vibration-space vector, respectively, j is $\sqrt{-1}$ and $x_n(t)$ is white noise. R_x is the autocorrelation matrix of the noisy signal x, which is the sum of the autocorrelation matrices of the signal x_s and the noise x_n given as follows:

$$R_x = R_{xs} + R_{xn} = SBS^H + \sigma_n^2 I = \sum_{k=1}^{P} I_k^2 s_k s_k^H + \sigma_n^2 I \tag{1.22}$$

where P is the number of frequencies, H represents the exponent of Hermitian transpose, I is the identity matrix, and s_k^H is the signal vector of complex sinusoids given by:

$$s_k^H = \left[1 \ e^{-j2\pi F_k} \ldots e^{-j2\pi F_k(N-1)}\right] \tag{1.23}$$

where $F_k = f_k/f_s$ in which f_s is the sampling frequency. B defines the power of the vibration space vector as $B = \text{diag}\,[I_1^2 I_2^2 \ldots I_p^2]$.

Finally, the pseudospectrum of MUSIC, Q, is defined as:

$$Q_i^{\text{MUSIC}}(F) = \frac{1}{\sum_{k=P+1}^{N} \left| s^H(F)\eta_k \right|^2} \tag{1.24}$$

where η is the eigenvector of the matrix R_{xs}, and $s_k^H(F_k)\eta_k = 0$ at the peaks of the frequency of the principal sinusoidal component.

Further details connecting the mathematical expression in Equations (1.21)–(1.24) are available in [177,283].

In this study to enhance the comparison among large pseudospectrum ranges the $Q_i(F)$, values are converted to logarithm base 10.

1.6.10 Feature extraction process

To identify the differences among the vibration responses of the different states of the structure, in addition to the energy feature of the signal components, four statistical time-domain features including RMS, shape factor, kurtosis, and entropy are extracted.

According to the literature, with any abrupt changes in the intensity of the vibration signal in various case studies, the values of RMS and energy features of the signal increase [224]. Therefore, the RMS and energy could be proposed to establish an effective indicator in identifying the damage.

The energy (E) of a given signal, $x(t)$, decomposed by EWT can be calculated from the sum of the square of the signal data as follows:

$$E = \int_0^t x(t)\, dt \tag{1.25}$$

Among the statistical features, the RMS is usually used to quantify the difference values between two vibrational signals. According to the literature [2], it has been proven that the value of RMS increases in proportion to the progress of a system fault or damage. The RMS of a given signal x_i (i.e., the sampled vibration signal) with N number of data points can be calculated as follows:

$$\text{RMS} = \sqrt{\frac{1}{N}\sum_{i=1}^{N} x_i^2} \tag{1.26}$$

As a non-dimensional statistical feature, shape factor (SF) of a signal can be defined from the ratio of RMS to mean of the sampled signal data as follows:

$$\mathrm{SF} = \frac{\sqrt{\frac{1}{N}\sum_{i=1}^{N} x_i^2}}{\frac{1}{N}\sum_{i=1}^{N} |x_i|} \tag{1.27}$$

Some advanced statistical-based features such as kurtosis and entropy features are calculated from the PDF of the vibration signal. PDF is a statistical expression that describes a probability distribution of each member of a discrete set of values of a variable or range of outcomes. According to the literature [2], it is obvious that the occurrence of any changes in vibration signals and stiffness of the structure causes the change of PDF. Consequently, the kurtosis and entropy features would also be affected and used to establish an efficient statistical test for identifying abrupt changes in the response of structures. In particular, kurtosis and entropy calculate the peak value and the histogram of the PDF from the vibration signal, respectively. As such, since the kurtosis is obtained from the peak of the PDF of the vibration signal [402], its value increases with any changes that occur in the PDF of vibration responses due to the appearance of damage.

The kurtosis (Ku) feature which is more effective in analyzing the non-stationary signals can be calculated from the peak values of the PDF of a signal given as follows:

$$\mathrm{Ku} = \frac{\sum_{i=1}^{N} (x_i - m)^4}{(N-1)\sigma^4} \tag{1.28}$$

where m and σ denote the mean and standard deviation of the sampled vibration signal.

In addition, various damage detection and SHM-based studies demonstrated the increment of entropy value with increasing the complexity of vibration response of structures that indicate the existence of damage [403,404]. The entropy, $e\,(p)$, is known as another statistical feature that is dependent on the PDF of a vibration signal. This feature is calculated based on the histogram of the PDF and represents the uncertainty and the degree of randomness degree of a sampled signal given as follows:

$$e(p) = -\sum_{i=1}^{N} p(x_i)\log_2 p(x_i) \tag{1.29}$$

where $p\,(x_i)$ denotes the normalized histogram of the sampled signal x_i.

1.7 STRUCTURE OF THE BOOK

SHM has become an important and hot topic for decades in various fields of civil, mechanical, automotive, and aerospace engineering, etc. Estimating the health condition and understanding the unique characteristics of structures through assessing measured physical parameters in real-time is the major objective of SHM. As a result, signal processing becomes an essential and inseparable approach of vibration-based SHM research. The basic goal of using signal processing is to identify the changes or damages from the vibration signals of the dynamic system to detect, locate, and quantify any damages existing in the system. Besides, Vibrations of complex structures such as bridges mostly present nonlinear and nonstationary behaviors. One of the serious issues of traditional signal processing techniques in analyzing the responses of real-life structures is related to the presentation of fundamental information of nonlinear, nonstationary, and noisy signals with closely spaced frequencies. To overcome this difficulty, numerous studies have been recently done to explore proper time-frequency signal processing techniques to efficiently present high-resolution representations for nonlinear characteristics of analyzed signals. Despite existing extensive reviews on vibration-based signal processing techniques in time and frequency domains for SHM purposes, there exists no study in categorizing the signal processing techniques based on the feature extraction in time, frequency and time-frequency representations. To fill this gap, firstly this book presents a comprehensive state-of-the-art review on the applications of time, frequency and time-frequency signal processing techniques for damage detection, localization, and quantification in various structural systems in Chapter 1. In addition, the theoretical background overview of the relevant specific signal processing approaches adopted in the present work to analyze the experimental measurements. In addition, a series of damage indices based on the signal features extracted by various signal processing techniques are proposed to evaluate the sensitivity in damage identification. Moreover, the significant and main objectives of this research are described in this section.

The research methodology for the implementation of combined signal processing techniques and ML in damage identification, localization, and quantification is described in Chapter 2. Moreover, in this chapter, the case study model and the experimental setup are introduced in detail.

In Chapter 3, the application of HHT based on three decomposition methods including EMD, EEMD, and CEEMDAN is investigated and compared in the damage detection process. Furthermore, the performance and of the sensitivity of different approaches such as MUSIC and EWT techniques in damage detection and quantification are assessed.

In Chapter 4, the experimental verification of the proposed damage detection using EMD-based techniques and their comparison, HHT, and MUSIC techniques are investigated in damage localization of the truss bridge model.

In Chapter 5, the efficiency of ANN is evaluated to automate the damage identification approaches with the purpose of effectively reducing the human intervention and speeding up the process of structure diagnosis.

Finally, Chapter 6 presents particular conclusions and main findings of this book, and also recommendation for future works.

Chapter 2

Methodology and approaches

2.1 INTRODUCTION

Owing to the difficulties of analyzing nonlinear systems, most of the previous research works considered the linear damage detection scenarios. Despite assuming linear damage scenarios in most of SHM studies of different civil engineering structures for many years, it has been proven that the dynamic behavior of the structure will differ from that of the undamaged model when the damage occurs and progresses in a structure [401]. The vast majority of previous studies that assumed structures as linear systems before and after damage, used the traditional modal techniques that are usually based on classical linear theory. However, these traditional techniques are not promising for SHM of complex structures since they are not efficiently sensitive to localized damage boundary conditions, sensor locations, and other environmental effects due to their linearity assumptions.

Generally, conventional signal processing techniques such as Fast Fourier Transform (FFT) [109,382,405], and Short-Time Fourier Transform (STFT) [174,212] are employed to analyze the linear and stationary signals. However, the responses of structures in the real world are inherently nonlinear and non-stationary. Despite higher extraction efficiency of the classical FFT compared to other algorithms, its deficiencies have previously been demonstrated associated with analyzing nonlinear and non-stationary signals. Although the STFT algorithm lacks multi-resolution analysis capabilities, it is able to analyze non-stationary signals since its theory is based on the windowed Fourier transforms. Due to some key advantages of frequency-domain methods on the basis of the Fourier transformation such as reducing the data volume and repairing the lost information by averaging the effects of random noise, these methods have been widely utilized in the literature. Unlike STFT, wavelet transform (WT) technique is a powerful tool allowing the incorporation of variable window sizes and capabilities in resolution and balancing both time and frequency domains of signals. However, similar to STFT, WT can only be used to process the signals under non-stationary and linear conditions.

DOI: 10.1201/9781003499046-2

To overcome the aforementioned drawbacks of frequency-domain traditional methods associated with linearity assumptions, a signal processing approach named Hilbert-Huang Transform (HHT) proposed by Huang et al. [406,407] which is promising in the analysis of nonlinear and non-stationary signals. The need for adopting HHT has been widely pronounced which is able to perfectly evaluate the nonlinear and non-stationary vibrations. The procedure of this technique contains two steps: (i) to decompose the original signal by the empirical mode decomposition (EMD) method into a series of complete and oscillatory components, named intrinsic mode functions (IMFs), (ii) to capture the instantaneous frequency and amplitude features by applying Hilbert transform (HT) to the IMFs. The vast majority of studies in the literature utilized HHT based on EMD for SHM and damage detection purposes [236,243,247,394,399,408,409]. Many research works in recent years have used HHT and EMD-based signal processing techniques [410–416] which do not have limitations of stationary and linearity. Indeed, the EMD incorporated the Hilbert spectrum method to identify the dynamic characteristics of linear structures. There exist many research works in the literature employing HHT, EMD-based, and WT-based [248,410,411] techniques for damage detection and localization of both linear and nonlinear vibration systems (particularly bridges) in which damage scenarios were designed by stiffness loss of truss bridge components or joints. Because of adaptively of EMD-based and HHT techniques, these techniques have been widely used for both linear and nonlinear vibration systems for many years [417–420]. The main reason for the use of EMD-based and HHT techniques in this study is that these techniques not only are adaptive with both linear and nonlinear vibration systems [417] but also they can localize any dynamic changes in time and frequency-based signal features of signal IMFs such as instantaneous amplitudes and instantaneous frequencies.

Despite major applications of EMD as a powerful signal processing tool in the literature, it has a significant inherent defect called mode mixing due to the intermittent frequencies in the IMF. To overcome the mode mixing problem of EMD, an improved EMD-based method proposed by Wu and Huang [271] named the ensemble empirical mode decomposition (EEMD) by adding an identical white Gaussian noise with an appropriate scale and standard deviation to the algorithm of EMD. However, the EEMD has two major drawbacks including the existence of residual noise in IMF, and its difficult procedure in computing the average of IMFs made by adding different white Gaussian noises to the analyzed signal. In order to solve the deficiencies of EEMD, Torres et al. [273] proposed CEEMDAN in which a particular noise is added at each stage instead of adding the white Gaussian noise which results in a unique residue after each IMF extraction. This approach also provides a proper spectral separation, increasing the level, and reducing the time of decomposition. Despite aforementioned improvements in the HHT technique, there are still some challenges in extracting close-spaced modes and noise susceptibility. Therefore, a more powerful technique with an ability to detect and extract the close-mode parameters of a structure with greater accuracy and noise immunity is of interest. In this chapter, the damage detection methodologies proposed in this book for damage identification, localization, and classification are introduced. Also, the methodologies from the combinations of novel signal processing techniques with artificial neural networks (ANN) are explained in details. Although CEEMDAN has been widely employed in mechanical [278,279,421–423], geological, seismic signals [424], and electrocardiogram (ECG) [273] fields, this technique has not been yet used for damage identification of civil engineering structures. Therefore, the motivation for using the CEEMDAN in this study is to assess the potential advantages of this procedure in damage detection of a laboratory-scale model of a steel truss bridge using a series of the conventional and improved damage indices in comparison with the performances of the previous generations of EMD-based techniques.

As well as evaluating the performance of CEEMDAN as a novel time-domain signal processing method, the capability of the empirical wavelet transform (EWT) as a time-frequency-domain technique proposed by Gilles [425] is also evaluated in this study in damage identification, localization, and classification of a steel truss bridge model. This technique can solves the drawbacks of traditional signal processing techniques with time-frequency representation such as inability of STFT in presenting the instantaneous frequencies with high-resolution, and the deficiency of WT in decomposing a time signal and calculating the instantaneous frequencies owing to incompatibility between the prescribed basis of WT and the analyzed signal, and mode mixing problem of EMD. The EWT is a popular adaptive decomposition algorithm that has an excellent capability in analyzing non-linear and non-stationary signals. In this method, the nonlinear signals are efficiently represented by time-frequency functions and instantaneous information of the analyzed signal can be adequately obtained in both time and frequency domains. The main advantages of utilizing EWT are to improve the frequency resolution limitations of data associated with the previous techniques, and efficiently separating the signal components in which the ratio of the signal low frequency to its high frequency is greater than 0.75. This technique has been widely used in the literature to accurately extract the components of the Fourier spectrum of signals.

Besides, the performance of a hybrid damage detection approach from the combination of CEEMDAN and multiple signal classification (MUSIC) named CEEMDAN-MUSIC as an improved frequency-domain technique is evaluated. The main purpose of such combination is the capability of the MUSIC algorithm which is able to provide a high-resolution power density spectrum of the signals decomposed by CEEMDAN in which the analyzed signal modes may be have very close frequencies and undistinguishable visually. This technique can effectively improve the detectability of close frequencies in comparison with the performances of traditional techniques such as FFT in analyzing signals with low signal-to-noise ratios (SNRs) [426].

One of the significant issues in vibration-based structural damage detection is to construct and extract the sensitive parameters from structural dynamic responses to identify the initial damages. Since structural health monitoring of complex civil structures such as bridges in which damage detection and localization are confusing procedures due to existing big data for analysis, a vital need for automated structural damage assessment systems to undertake the soundness, validity, and efficiency of the diagnostic process, is the topic of importance. To this end, multifarious structural damage identification approaches based on the combination of signal processing techniques including CEEMDAN and EWT and Artificial Neural Network (ANN) technique has been developed in this study which can guarantee the robustness, reliability, and efficiency of the detection process.

The main advantage of ANNs compared to other machine learning algorithms is that they allow the nonlinear and non-parametric modelling of complex systems with large datasets simply and adaptively without requiring explicit mathematical representations or without requiring exhaustive experiments. However, most other statistical methods are parametric models that need a high background in statistics. Hence, the use of ANN and its hybridization with signal processing techniques are more popular for structural health monitoring (SHM) applications compared to other intelligent algorithms. Therefore, this study proposes a series of automated damage detection approach from the combination of improved signal processing technique including CEEMDAN and EWT techniques with ANN technique (named CEEMAN-HT-ANN and EWT-ANN techniques) based on several time-dependent features including RMS, shape factor, kurtosis, and entropy and statistical features, which is the research gap of previous works. In addition, a series of damage indices based on three signal features including the energy, instantaneous amplitude (*IA*), and unwrapped phase (*P*), and four statistical time-domain features including RMS, shape factor, kurtosis, and entropy. In addition, four improved damaged indices based on the combinations of kurtosis (*Ku*) and entropy (*e*) features with each of the energy and *IA* features are proposed.

In this book, the applications of the most recent time-frequency signal processing techniques and as well as their improved versions are evaluated for structural damage detection, quantification and localization in a scaled model of a steel truss bridge when subjected to different excitations. The main contributions of this study is listed as follows:

- As discussed in Section 2.1 and reviewed in Chapter 1, the accuracy and superior performances of CEEMDAN and EWT techniques as novel and improved time-frequency signal processing techniques have been proven in the literature in comparison with the traditional sole frequency- and time-domain techniques. Because these methods are efficiently able to analyze noisy signals with nonlinear characteristics and capture the fundamental information of analyzed signals with high-resolution representations in both time and frequency domains. Hence, they can accurately identify even invisible damages occurred in the internal section of the structures through the variation of non-linear structural characteristics. The advantages of the CEEMDAN technique in damage detection and quantification compared to EMD and EEMD techniques are verified in Section 2.11.

- In addition, since the application of improved techniques on complex civil engineering structures with big data acquired would require high computational cost, a series of improved damage detection approaches have been proposed based on the combinations of CEEMDAN and EWT techniques with ANN to automate the damage detection procedures in a reliable, quick, and simple manner for a complex truss bridge structure. This automation provides an intelligent decision-making procedure to accurately detect the presence and location of the damage in complex civil structures such as truss bridges in which data is quite big. The accuracy of the proposed frameworks is assessed by quantifying the sensitivities of several time-dependent signal features including energy, instantaneous amplitude (*IA*), and unwrapped phase (*P*), and four statistical time-domain features including RMS, shape factor, kurtosis, and entropy. The procedures of the proposed approaches based on the combinations of CEEMDA and EWT with ANN have been described in detail in Sections 2.4 and 2.5, respectively.
- In addition, the hybrid use of MUSIC and CEEMDAN techniques for damage identification of a steel truss bridge is proposed in this study. This proposal aims to resolve the limitation of CEEMDAN in selecting the signal noise ratio (SNR) and notable overlapping that may be generated between the modes when analyzing noisy signals. Because MUSIC algorithm as the most improved frequency-based technique can efficiently improve the detectability of close frequencies of noisy signals with low SNRs. The procedure of the proposed approach based on the combination of CEEMDAN and MUSIC has been described in Section 2.6. In addition, the advantage of the proposed CEEMDAN-MUSIC compared to pure MUSIC technique is verified in Section 2.12.
- As another research contribution of this book, a series of damage indices based on the combination of advanced statistical-based signal features including kurtosis and entropy features and energy and *IA* features. These damage indices works based on the application of kurtosis and entropy functions (which are sensitive to probability density function (PDF) of the vibration signals) on two main signal features of the analyzed signal by CEEMDAN including energy and *IA* features instead of their applications on the original signal. This is because of the fact that the PDF-based interpretation of signal features extracted by improved methods such as CEEMDAN which are perfectly able to reconstruct the low-SNR signals, can more accurately address the information of damages and faults. The details of the proposed damage indices are given in Section 2.7.

2.2 TARGET OF INVESTIGATION

A laboratory-scale model of a steel truss bridge with fourteen bays, a span length of 5.6 m, and a height of 0.4 m built in the structural engineering laboratory of the Qingdao University of Technology (QUT), as shown in Figure 2.1, is studied based on the design characteristics of the truss bridge model established in the University of Illinois at Urbana-Champaign [427]. The length of each horizontal and vertical element is 0.4 m, and the length of each diagonal element is $0.4\sqrt{2}$ m. The bridge has a roller support at one end and a hinge support at the other end as shown in Figure 2.2. All the elements of the bridge in its healthy state are steel tubes with inner and outer diameters of 1.09 cm and 1.71 cm, respectively. The details of an example accelerometer sensor attached to a typical joint which connects the elements to each other through screwing are illustrated in Figure 2.2. A band-limited white noise excitation with an RMS of 0.4 and a bandwidth of 1.0 Hz as shown in Figure 2.3a is developed in MATLAB and then, it is applied to the bridge using a shaker device (CF6900-100) with the maximum force-generating capacity of 100 N which is able to generate vibrations with frequencies between 10 Hz and 2500 Hz. The magnitude of the excitation generated by the shaker is controlled using a power amplifier device

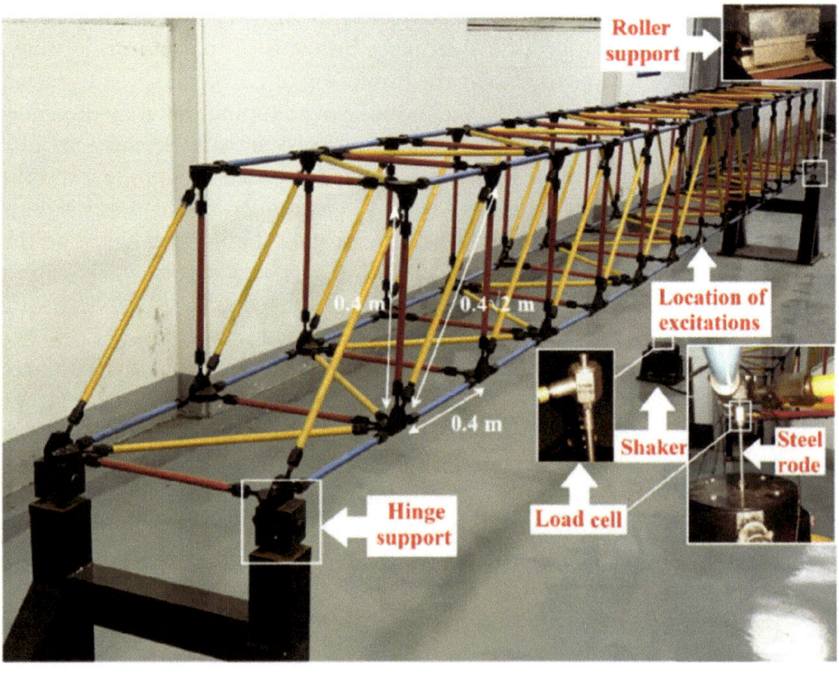

Figure 2.1 A perspective view of the steel truss bridge model as the case study in the present work established in the QUT laboratory.

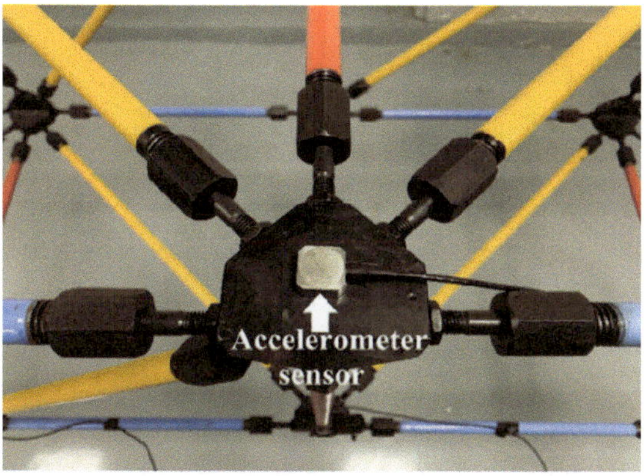

Figure 2.2 Details of a typical joint with the attached acceleration sensor.

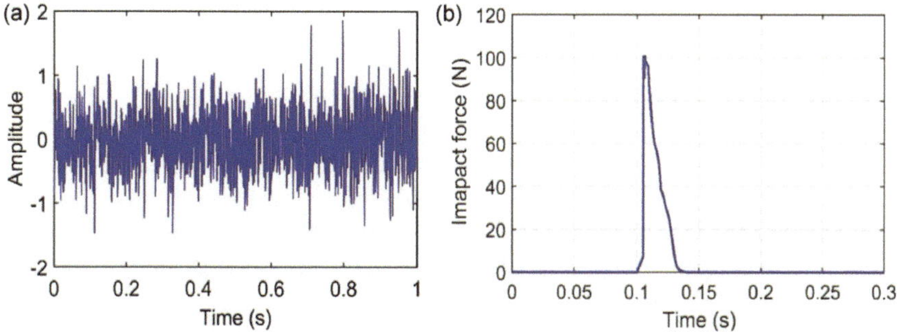

Figure 2.3 Excitations applied to the truss bridge mode; (a) a band-limited white noise by the shaker, (b) an impact impulse by a typical hand-held impact hammer.

(CF6502) and it is applied to the specified joint on the truss through a steel rode with an attached load cell (CF3110) with a sensitivity of 25.32 mV/N which measures the magnitude of the applied force.

In addition, the impact excitations are applied to the bridge using a typical hand-held impact hammer. The frequency content of the input impact force and consequently the input spectrum is extremely dependent on the stiffness, mass, and velocity of the struck hammer. In this study, an impact impulse with a peak input force of 100 N, and a duration about 50 ms is applied to the bridge as shown in Figure 2.3b. This impact load represents an impulsive loading type since the duration of the impact load equal to 35 ms is less than the first natural period of the truss bridge in its healthy state which is 51 ms (with the corresponding first natural frequency of 19.53 Hz) [224]. In order to measure the vibration responses of the truss bridge, a series of piezotronic accelerometers (herein, thirteen pieces) piezotronic accelerometer sensors (CF0420) with a sensitivity of mV/g are uniformly distributed and attached to throughout the lower chord of the bridge

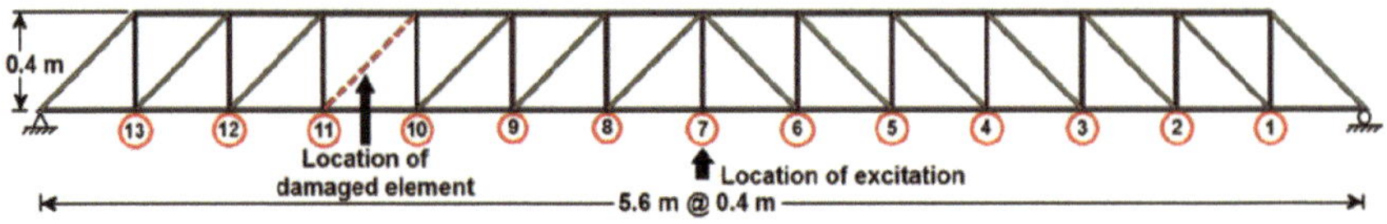

Figure 2.4 A schematic view of the truss bridge model by showing the uniform distribution of the locations of sensors, and the location of the damaged element.

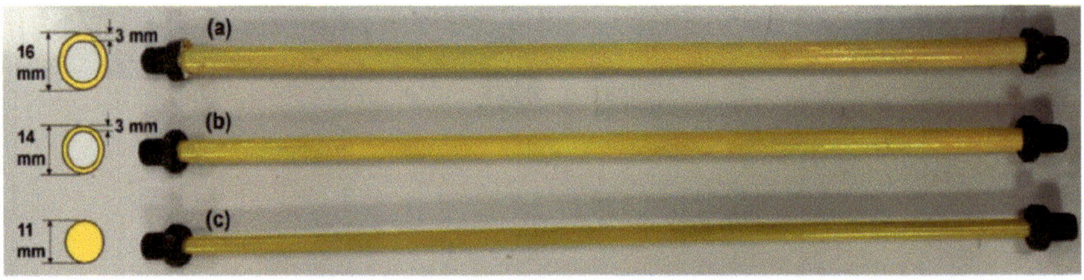

Figure 2.5 Details of the damaged elements with; (a) 20%, (b) 50% and (c) 80% reduced percentages of the cross-section stiffness.

span as shown in Figure 2.4. Then, the data acquisition system (CF3820) records the measured responses with a sampling frequency of 2000 points per second through eight channels.

In this study, four damage scenarios are integrated for the truss bridge. Three of these scenarios are implemented by replacing a diagonal element located between the locations of sensors 10 and 11 as illustrated in Figure 2.3. The severities of three damaged states of the bridge are accomplished reducing by 35%, 60%, and 83% of the cross-sectional stiffness of the specified element (i.e., the specified diagonal element in Figure 2.4) as shown in Figure 2.5. The reductions of the cross-sectional moment of inertia of these damaged elements in percentage terms are; (a) 35% in which the outer and inner diameters are 16 mm and 10 mm, respectively, (b) 60% in which the outer and inner diameters are 14 mm and 8 mm, respectively, (c) 83% in which the diameter of the bar element is 11 mm. Besides, the 100% damaged state of the bridge is designed by removing the specified diagonal element. The axial stiffness reductions of these damaged elements are 14%, 27%, and 33%, respectively. It is noteworthy that excluding the aforementioned diagonal element, other structural characteristics of the bridge and the loading excitations are kept constant for all the healthy and damaged scenarios.

2.3 PROCEDURE OF CEEMDAN-HHT-ANN DAMAGE DETECTION METHOD

An efficient damage detection approach is proposed based on vibration signal processing methods. In order to evaluate the performance of the CEEMDAN-based damage detection approach, first, the vibrations at the identified accelerometer sensors of the steel truss bridge model when subjected to a band-limited white noise are acquired in the healthy and damaged states. Afterwards, the vibrations are decomposed using the CEEMDAN technique to generate the set of IMFs. Then, the HT is applied to the first IMF to extract three features including energy, unwrapped phase, *IA*. In addition, four improved damaged indices based on the combinations of kurtosis (*Ku*) and entropy (*e*) features with each of the energy and *IA* features are proposed. After employing the HHT, a multi-layer perceptron (MLP) neural network is defined to train the relationship between IMFs as the input layers and their seven aforementioned features for the healthy state of the bridge as the output layers as shown in Figure 2.6. The ANN is separately trained based on each feature before applying the damage scenarios on the bridge. The outputs of the CEEMDAN-HT-ANN model based on seven aforementioned features extracted from the healthy state of the bridge are compared with those from the scenarios with the variation of the damage level, and the sensor location relative to the damaged element to classify the severity and detect the location of damage, respectively. Finally, the defined damage indices and HHT spectrums are adopted to identify the presence, location, and severity of the damage in the truss. The flowchart diagram of the proposed methodology is shown in Figure 2.6.

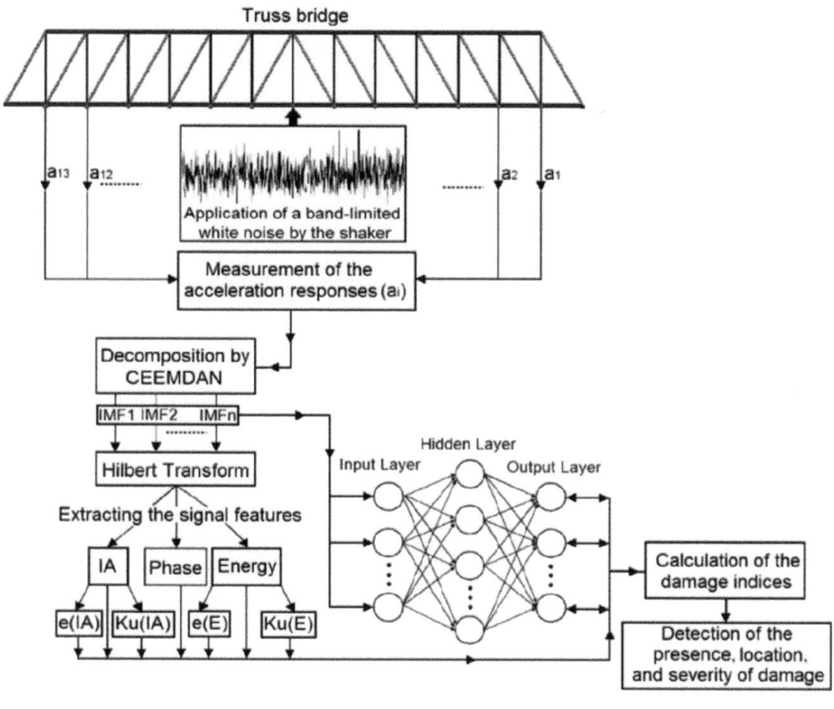

Figure 2.6 The framework of the proposed CEEMDAN-HT-ANN damage detection approach.

2.4 PROCEDURE OF EWT-ANN DAMAGE DETECTION METHOD

The procedure of the proposed damage detection approach named EWT-ANN is introduced in this section. In the first step of this method, the vibration signals of the steel truss bridge in healthy and damaged states are acquired by the accelerometer sensors when subjected to band-limited white noise and impact excitations. Then, the EWT is used to decompose the original signal into several modulated signal components and modes. These signal components are defined as the inputs to the ANN. Thereafter, the statistical time-domain features including RMS, shape factor, kurtosis, and entropy are extracted from the generated signal modes to define them as the outputs to the ANN. The ANN trains relationships between the signal components as the input layers and the extracted features as the output layers through a multi-layer perceptron (MLP) neural network for healthy the healthy state of the bridge. The outputs of the trained EWT-ANN procedure on the healthy state of the truss is considered as the target. Besides, the ANN is separately applied to the sketched damaged scenarios of the bridge. Finally, the outputs of the EWT-ANN approach calculated based on the aforementioned features for the damaged states of the bridge are compared with the target (i.e., the outputs of healthy sate) to assess the performance of the proposed model in detecting and locating the damage. The framework flowchart of the proposed approach is shown in Figure 2.7 and further explanations of this flowchart can be found in Chapter 4.

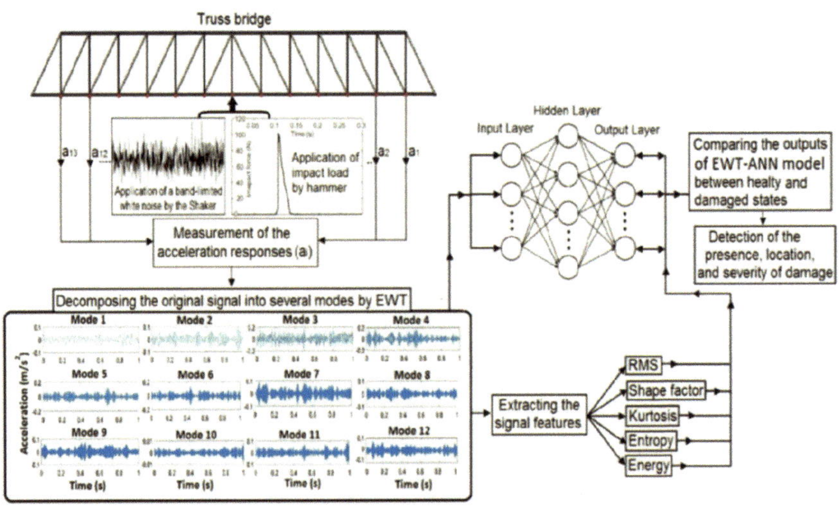

Figure 2.7 The framework of the proposed EWT-ANN damage detection approach.

2.5 PROCEDURE OF CEEMDAN-MUSIC DAMAGE DETECTION METHOD

In this section, an efficient damage detection approach is introduced based on the combination of signal processing methods. In order to evaluate the performance of the combined CEEMDAN-MUSIC technique, first, the vibrations from the identified accelerometer sensors on the steel truss bridge model are acquired when subjected to a band-limited white noise in the healthy and damaged states. Then, the vibrations are decomposed using the CEEMDAN method to generate the set of IMFs. Afterwards, the MUSIC algorithm is applied to the IMFs to transform the time series acceleration response to the power density spectrum domain (i.e., pseudospectrum). It is noteworthy that in this study, the MUSIC algorithm is applied to only IMF1. Finally, the absolute differences between the pseudospectral peaks in terms of the frequency (D_f) (normalized using the pseudospectral peak from the healthy case) and the power magnitude (D_p) ranges are considered as the damage indicators. By analyzing these damage indicators, the presence, severity, and location etc., namely the characterization of damages are determined. The performance of the proposed combined CEEMDAN-MUSIC technique is assessed by comparing the damage indicator results with those from the pure MUSIC algorithm. The framework flowchart of the proposed methodology is shown in Figure 2.8 and further explanations of this flowchart can be found in Chapter 4.

Although CEEMDAN significantly reduces the computational cost by decreasing the number of useless IMFs generated in EMD and EEMD

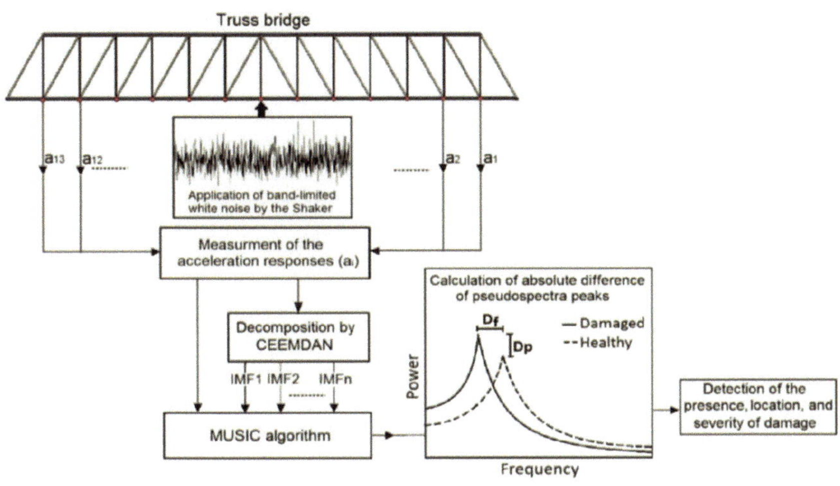

Figure 2.8 The framework of the proposed methodology.

algorithms, it has a limitation in selecting the signal noise ratio (SNR) which is extremely dependent on the characteristics of the analyzed signal. As such, during the decomposition process of a signal by CEEMDAN, notable overlapping may be generated between the modes due to adding a particular EMD mode of white noise with different signal noise ratios (SNR) at each step.

Therefore, an empirical approach using trial-and-error technique is needed to select an efficient coefficient of SNR to analyze a particular signal. Besides, the implementation process of the proposed methodology in this paper which assesses a scaled model of a truss bridge built in a laboratory environment has several limitations compared to those employed in real environment listed as follows:

- Implementing a preprocessing step to denoise the signals of a noisy environment.
- Considering the effect of various factors on the vibrational measurements such as the evolution of boundary condition and ambient condition (e.g., temperature, traffic, etc.)
- Utilizing artificial intelligent methods to classify the huge amounts of data generated by big and complex structures.

2.6 DAMAGE INDICES

In this study, a series of damage indices based on three signal features including the energy, instantaneous amplitude (*IA*), and unwrapped phase (*P*), and four statistical time-domain features including RMS, shape factor, kurtosis, and entropy extracted by EMD, EEMD, and CEEMDAN techniques, are used to detect, locate, and classify the severity of the damage in a laboratory-scaled model of a steel truss bridge. These damage indices compare the differences between the characteristics of the healthy and damaged states.

The damage index based on the energy feature has been widely used in the literature [420,428]. Cheraghi et al. [429,430] introduced a damage index based on the energy of the first IMF of the vibration signals. In order to obtain the energy of each IMF, the vibration of the structure in its healthy and damaged states are captured and decomposed by CEEMDAN. Then, the energy of the respected IMF for each sensor is calculated according to Equation (2-1):

$$E = \int_{0}^{t_0} (IMF)^2 dt \tag{2.1}$$

where the value of parameter E is a scalar. Finally, the energy damage index is defined as:

$$DI_{(E)} = \left| \frac{E_{\text{Healthy}} - E_{\text{Damaged}}}{E_{\text{Healthy}}} \right| \times 100 \tag{2.2}$$

The high scalar value of *DI* represents the existence, location, and severity of damage in the structure based on identified scenarios in the following sections. The same procedure is repeated for the *IA* parameter. However, this feature is extracted by applying the Hilbert transform (HT) to each IMF and calculating the average of the desired IMF.

$$\overline{IA} = \frac{1}{n}\sum_{i=1}^{n} IA_i \tag{2.3}$$

where $\overline{IA}$ represents the mean of the instantaneous amplitude of each IMF, and n is the number of time samples for $IA_i=[IA_1,IA_2,\ldots,IA_n]$. Accordingly, the damage index based on *IA* can be defined as follows:

$$DI_{(IA)} = \left|\frac{\overline{IA}_{\text{Healthy}} - \overline{IA}_{\text{Damaged}}}{\overline{IA}_{\text{Healthy}}}\right| \times 100 \tag{2.4}$$

According to the literature [406], the calculation procedure of the unwrapped phase parameter (*P*) is based on the averaging of the IMFs' phases produced by applying a Hilbert transform (HT) to each IMF given as follows:

$$\overline{P} = \frac{1}{n}\sum_{i=1}^{n} P_i \tag{2.5}$$

Where $\overline{P}$ represents the mean unwrapped phase of each IMF extracted by HT. Accordingly, the damage index based on *P* can be defined as follows:

$$DI_{(P)} = \left|\frac{P_{\text{Healthy}} - P_{\text{Damaged}}}{P_{\text{Healthy}}}\right| \times 100 \tag{2.6}$$

Some advanced statistical-based features such as RMS, kurtosis and entropy features are calculated from the PDF of the vibration signal. PDF is a statistical expression that describes a probability distribution of each member of a discrete set of values of a variable or range of outcomes. According to the literature [2], it is obvious that the occurrence of any changes in vibration signals and stiffness of the structure causes the change of PDF. Consequently, the kurtosis and entropy features would also be affected and used to establish an efficient statistical test in identifying abrupt changes in the response of structures. In particular, kurtosis and entropy calculate the peak value and the histogram of the PDF from the vibration signal, respectively. As such, since the kurtosis is obtained from the peak of the PDF of the vibration signal [402], its value increases with any changes occurred in the PDF of vibration responses due to the appearance of damage.

Among the statistical features, the RMS is usually used to quantify the difference values between two vibrational signals. According to the literature [2], it has been proven that the value of RMS increases in proportion to the progress of a system fault or damage. The RMS of a given signal x_i (i.e., the sampled vibration signal) with N number of data points can be calculated as follows:

$$\mathrm{RMS} = \sqrt{\frac{1}{N}\sum_{i=1}^{N} x_i^2} \tag{2.7}$$

As a non-dimensional statistical feature, shape factor (SF) of a signal can be defined from the ratio of RMS to mean of the sampled signal data as follows:

$$\mathrm{SF} = \frac{\sqrt{\frac{1}{N}\sum_{i=1}^{N} x_i^2}}{\frac{1}{N}\sum_{i=1}^{N} |x_i|} \tag{2.8}$$

In addition, two advanced statistical-based features including kurtosis (Ku) and entropy (e) features that are calculated from the PDF of the vibration signal, are considered in this study. Furthermore, in order to extract more sensitive indices and improve the performance of the aforementioned damage indices based on the energy and IA features, the combinations of two statistical damage indices including kurtosis and entropy with the energy (E) and instantaneous amplitude (IA) features extracted using CEEMDAN techniques are proposed. To do this, instead of applying these statistical features to the original vibration signal, both of the kurtosis and entropy features are applied to the energy and IA time-histories extracted from the IMFs of the analyzed signal using CEEMDAN.

The kurtosis (Ku) feature which is more effective in analyzing the non-stationary signals can be calculated from the peak values of the PDF of a signal given as follows:

$$Ku = \frac{\sum_{i=1}^{N} (x_i - m)^4}{(N-1)\sigma^4} \tag{2.9}$$

where m and σ denote the mean and standard deviation of the sampled vibration signal. Therefore, the damage indices based on the combinations of the kurtosis with energy ($DI_{(Ku\ (E)}$), and with IA ($DI_{(Ku\ (IA))}$) features can be defined as given in Equations (2.10) and (2.11), respectively.

$$DI_{(Ku(E))} = \left| \frac{(Ku(E))_{\text{Healthy}} - (Ku(E))_{\text{Damaged}}}{(Ku(E))_{\text{Healthy}}} \right| \times 100 \tag{2.10}$$

$$DI_{(Ku(IA))} = \left| \frac{(Ku(\overline{IA}))_{\text{Healthy}} - (Ku(\overline{IA}))_{\text{Damaged}}}{(Ku(\overline{IA}))_{\text{Healthy}}} \right| \times 100 \tag{2.11}$$

In addition, various damage detection and SHM based studies demonstrated the increment of entropy value with increasing the complexity of the vibration response of structures that indicate the existence of damage [403,431]. The entropy (e) is known as another statistical feature that is dependent on the PDF of a vibration signal. This feature is calculated based on the histogram of the PDF and represents the uncertainty and the degree of randomness degree of a sampled signal given as follows:

$$e = -\sum_{i=1}^{N} p(x_i)\log_2 p(x_i) \tag{2.12}$$

where $p\ (x_i)$ denotes the normalized histogram of the sampled signal x_i. The damage indices based on the combinations of the entropy with energy ($DI_{(e(E)}$), and with IA ($DI_{(e(IA))}$) features can be defined as given in Equations (2.13) and (2.14), respectively.

$$DI_{(e(E))} = \left| \frac{(e(E))_{\text{Healthy}} - (e(E))_{\text{Damaged}}}{(e(E))_{\text{Healthy}}} \right| \times 100 \tag{2.13}$$

$$DI_{(e(IA))} = \left| \frac{(e(\overline{IA}))_{\text{Healthy}} - (e(\overline{IA}))_{\text{Damaged}}}{(e(\overline{IA}))_{\text{Healthy}}} \right| \times 100 \tag{2.14}$$

2.7 BENCHMARK STUDIES

This section presents a series of experimental benchmark studies of the signal processing techniques used in this book in which their applicability and advantages in damage detection of civil and structural engineering structures.

2.7.1 EMD-based HHT technique

The validity of EMD and HHT techniques were studied on a benchmark problem established by the ASCE Task Group on Structural Health

Monitoring [432]. The structure selected for the benchmark problem is the four-story 2-bay by 2-bay steel braced frame as depicted in Figure 2.9. The structure has a 2.5m×2.5m base, is 3.6m tall and is located at the Earthquake Engineering Research Laboratory of the University of British Columbia.

This benchmark problem has been studied using different approaches, for instance, the probabilistic based estimations [393,433], the eigensystem realization algorithm [434], a multistage approach for damage identification [435], etc. In these studies, the damage is detected by a comparison of the system properties of damaged and undamaged structures.

From the application of EMD-based HHT technique on the seismic acceleration response of the ASCE benchmark by Yang et al. [393], its capability in extracting the damage spikes due to a sudden change of structural stiffness from the measured data as shown in Figure 2.10 thereby detecting the damage

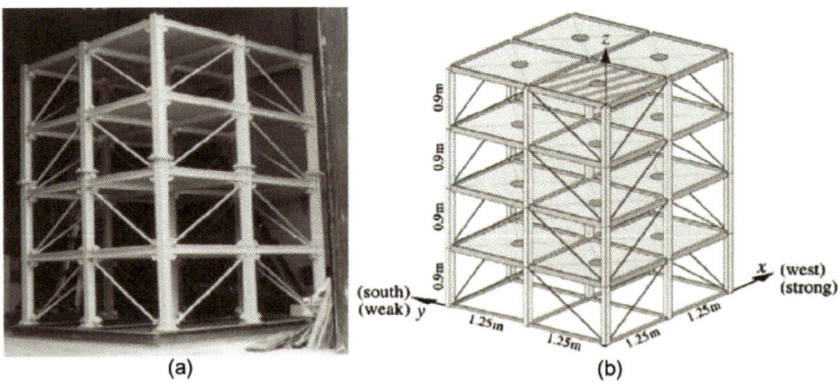

Figure 2.9 A four-story ASCE benchmark structure; (a) experimental test [48], (b) a schematic of the model [393].

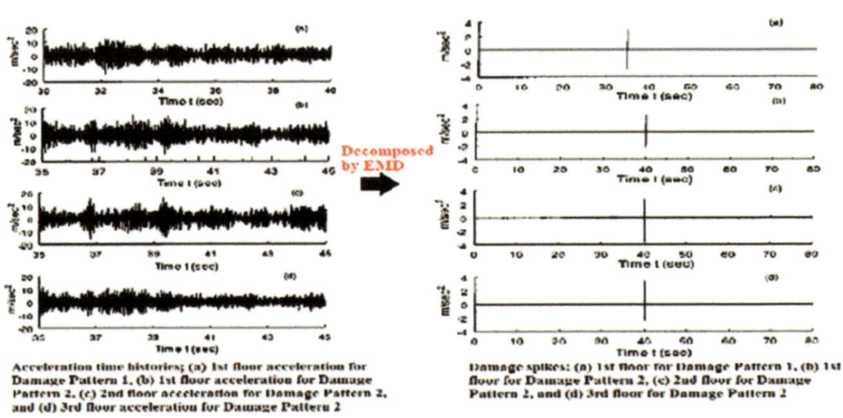

Figure 2.10 Application of EMD-based HHT on the acceleration response of the ASCE benchmark by Yang et al. [392].

time instants and damage locations as shown in Figure 2.11, was concluded. Also, the time instant when the damage occurs regardless of the noise level was accurately characterized as well as successfully identification the natural frequencies and damping ratios of the structure before and after damage.

The ASCE benchmark problem was later studied by Liu et al. [393] to test the applicability of EMD-based HHT. The results showed the capability of HHT method in: (a) recovering the actual signals 'time–frequency feature; (b) detecting and locating the actual structural damages; (c) detecting the instant of the impact load or of the damage taking place in active structural health monitoring as shown in Figure 2.12 In this method, the EMD was used to eliminate noise and it was found that that EMD method is better than the Wavelet method for the de-noising purpose.

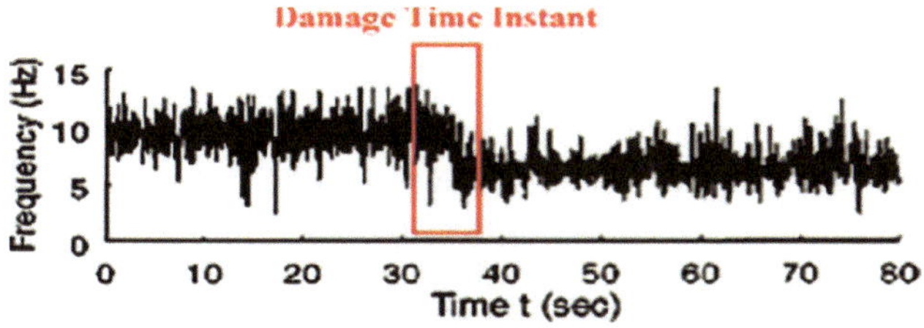

Figure 2.11 Frequency-time relation associated with maximum amplitude for the first modal response of the structure [392].

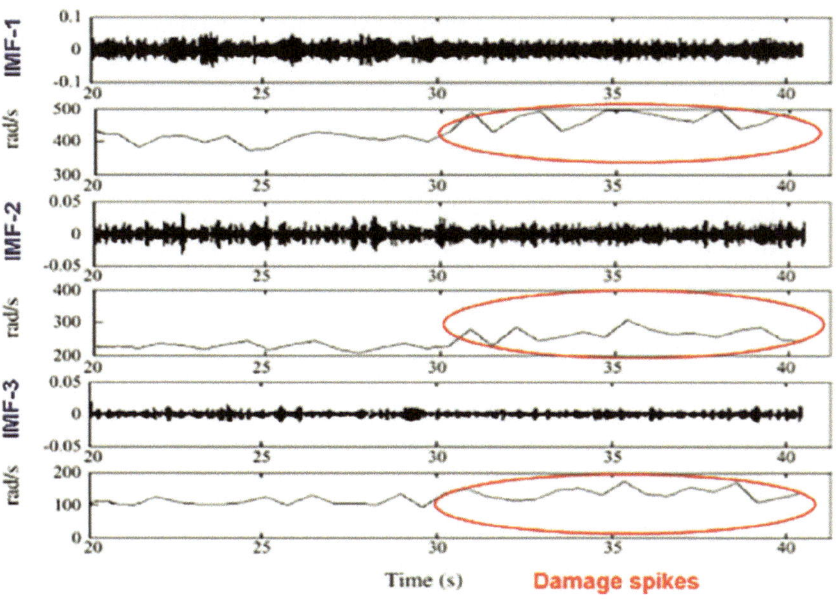

Figure 2.12 The constructed signal's IMFs and time–frequency [393].

Lin et al. [436] conducted the damage identification of structural health monitoring benchmark building by using a HHT-based technique for linear structures. The structural parameters, including the stiffness and damping, before and after damage were identified first and then the location and severity of the damage are assessed by a comparison. The made observations indicate that the accuracy of the HHT technique presented for identifying the structural damages is very plausible, and it represents a viable system identification and damage detection technique for linear structures. In addition, Taha [437] used the ASCE benchmark structure to demonstrate the ability of an integrated method based on using artificial neural network (ANN) in computing the wavelet energy feature of acceleration signals acquired from the structure.

2.7.2 CEEMDAN technique

A three-story frame model developed at the Los Alamos National Laboratory (LANL) [438] was used by Civera and Surace [439] as a benchmark problem to study the capability of the CEEMDAN technique considering nonlinear distortions, stiffness reductions, and added masses. The model excited horizontally at the moving base as shown in Figure 2.13. More detailed information about the LANL benchmark can be found in the previous studies [438,440].

From the application of the CEEMDAN, the following conclusions were obtained:

- Since the CEEMDAN algorithm shows a large variability due to its step involving the addition of random White Gaussian Noise (WGN), it was found that even with all the improvements with respect to the standard EMD, the CEEMDAN algorithm still suffers an over-decomposition issue. Also, only a few of the first IMFs could capture modes in the time and frequency domains for the linear baseline. That is, the first IMF included the second and the third eigenmodes of the structure, while the second IMF mostly captured the first mode. Thus, the algorithm failed to separate the linear normal modes (LNMs) of the system.
- Regarding the effects of filtering, removing the frequencies out of the input range was not particularly beneficial nor detrimental for the CEEMDAN algorithm. The number of the extracted IMFs was not sensibly reduced. The first modes were limitedly affected, while residual IMFs changed noticeably in both the frequency and the time domain; however, these latter components are of almost null utility for SHM purposes.
- The CEEMDAN-extracted IMFs were the only ones with noticeable (even if very small) fluctuations of the IF in their baseline model. The first IMF seems to be relatively sensitive to damage and structural changes, as it slightly deviates from the predicted value

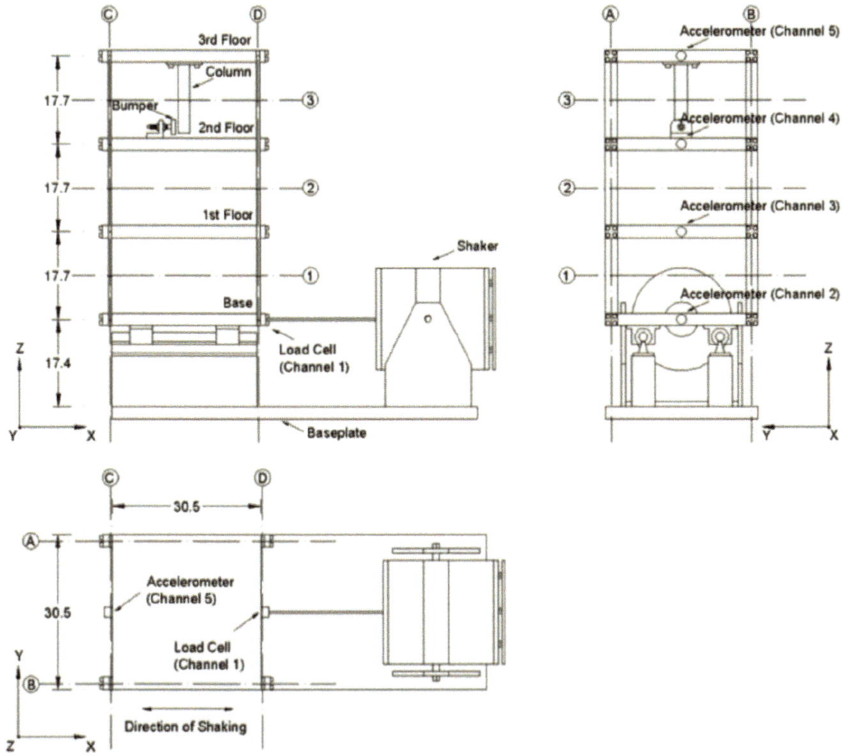

Figure 2.13 The experimental test setup schematics [438]; (a) Front view; (b) side view.

- The second and third IMFs seemed to be unchanged by the added mass or the inserted point wise source of nonlinearity. This can be both an advantage (e.g., for masonry structures, where many material and geometrical nonlinearities cause noise-like, damage-unrelated distortions) or a limitation (e.g., being insensitive to breathing cracks' effects). Nevertheless, such deviation from the normality model is arguably too small and time limited to signify damage with good confidence.
- Finally, the capability of the CEEMDAN algorithm in isolating the most important characteristics of the target signal was concluded while it was greatly affected by measurement noise and nonlinear distortions.

2.7.3 MUSIC technique

Jiang and Adeli [441] used the MUSIC technique for the first time in civil and structural engineering. The structural damage was identified explicitly through comparison of the pseudospectra of the output predicted by the trained model and the other based on output measured by the sensor. The validity of MUSIC was revealed using the experimental data obtained for a scaled model of a 38-story RC structure. The results showed that the proposed methodology provides an effective method for damage detection in high-rise building structures and a powerful tool for real-time health monitoring and non-destructive damage evaluation of civil structures.

2.8 VALIDITY OF SIGNAL FEATURES

In this section, the applications of several signal features and parameters used in this study are review to verify their validity in SHM and damage detection of various structural and mechanical systems.

From an experimental study of a single-span bridge under moving load excitation by Kunwar et al. [442] as shown in Figure 2.14, the presence and location of damage in the bridge was accurately identified by evaluating the instantaneous amplitude, phase, and frequency (i.e., the Hilbert spectrum) features extracted from the Hilbert transform of IMFs of the transient vibrations using EMD technique. From the results of a sensor close to damage location, significant increase in magnitudes throughout the frequency range, reduction in peak frequency of Hilbert spectrum, and considerable reduction in values of instantaneous phase were concluded compared with the baseline (healthy) bridge structure.

The capability of energy feature in measurement of uncertainty associated with damage detection in an experimental laboratory model structure concluded by Mohammadi Ghazi and Büyüköztürk [408]. In addition, the damage-induced nonlinearities in the structural response was effectively captured by the energy feature of the analyze signals (Figure 2.15).

The energy damage index calculated based on EMD technique was used by Esmaeel et al. [443] to establish the existence of damage due to loosened bolts in common industrial bolted joints in pipes transporting oil and gas as

Figure 2.14 Experimental bridge model with wheel and axle used for excitation [442].

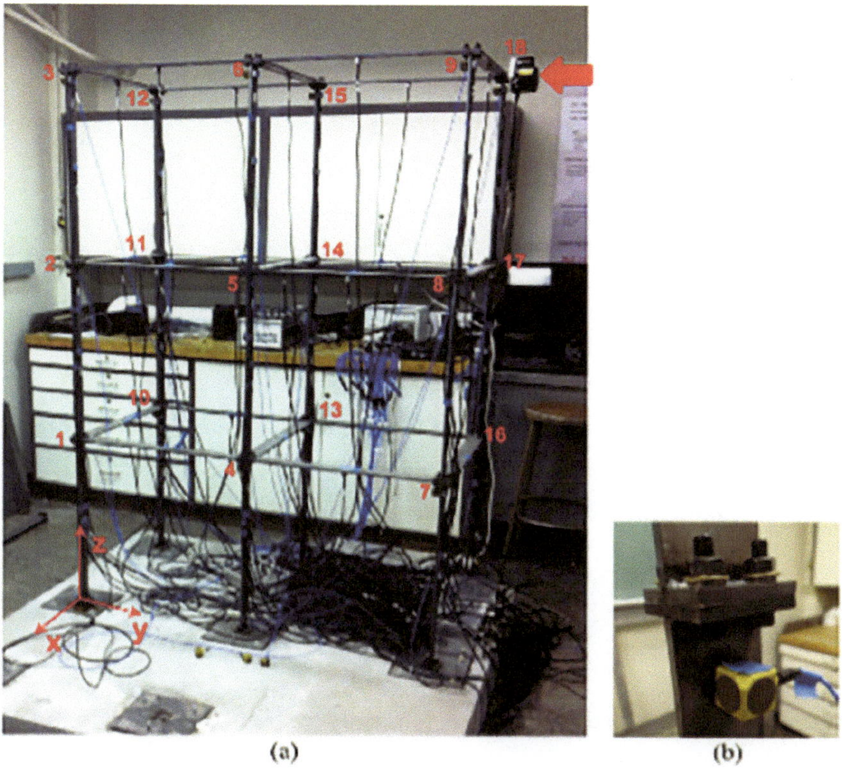

Figure 2.15 The experimental setup of: (a) the three-story two-bay structure and (b) a sensor next to a connection [408].

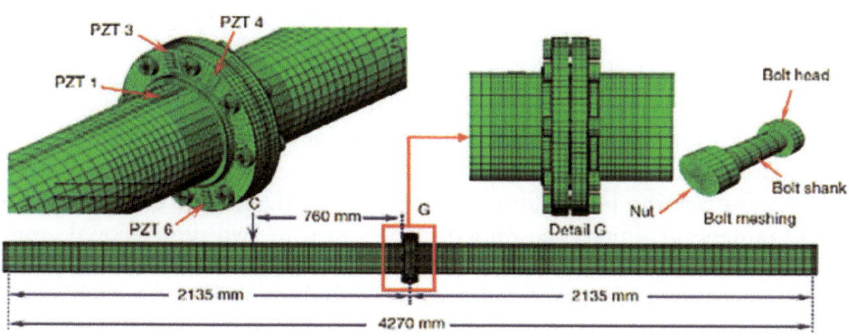

Figure 2.16 Details of bolted joints in pipes transporting oil and gas [443].

shown in Figure 2.16. It was found that the energy-based damage index is a powerful indicator for not only detecting the damage, but also the progression of the damage in bolted joints.

Figure 2.17 Laboratory model of a derrick steel structure [444].

Han et al. [444] revealed that the capability of instantaneous energy and amplitude curvature methods in accurately determination of the location of the damage element and weak damage element in a derrick steel structure as shown in Figure 2.17 as well as qualitatively analyzing the damage degree of the element; under the impact load, the noise hardly affects the identification of the damage location.

Tua et al. [445] used the energy peaks in the Hilbert spectrum corresponding to crack-reflected waves to determine accurate flight times and also to estimate the orientation of the crack.

Pines and Salvino [446] illustrated how the EMD method can be used to track damage in simple one-dimensional and thin two-dimensional composite structures. The study focused on developing and validating damage detection schemes based on the features of the EMD associated with Hilbert spectral analysis including energy density, the time of flight, and the energy speed propagation. It was found that the energy–time spectrum could locate the damage, whereas the energy density spectrum was used to quantify the damage.

Besides, a larger reflection due to a larger damage resulted in an increase of the energy in the Hilbert energy spectrum. Also, it was found that the instantaneous phase can be used as a damage detection tool. The Hilbert phase could not only detect the presence of the defect in the structure but

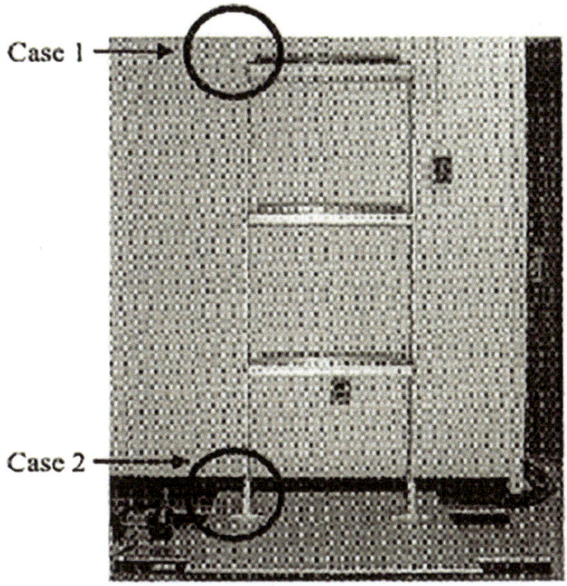

Figure 2.18 Three-story scale building model [448].

also quantify the extent of the damage. A Hilbert energy spectrum describing the wave energy density can be created and the severity of the damage inferred from the reflected energy. It was revealed that the amplitude of the reflection increased with the size of the damage leading to an increase in Hilbert energy spectrum. Therefore, the energy–time–frequency plot provided by the HHT could be used to identify and assess structural damage and a Hilbert energy spectrum describing the wave energy density can be created and the severity of the damage inferred from the reflected energy. In addition, the use of unwrapped instantaneous Hilbert phase feature has been suggested by previous studies [447] for damage detection metrics. The instantaneous frequency and phase features of vibration responses of a civil building model as shown in Figure 2.18 extracted by HHT technique coupled with EMD were used by Pines and Salvino [448] as the damage indicators for detecting the structural damage.

2.9 IMPLEMENTATION OF SIGNAL PROCESSING TECHNIQUES

2.9.1 HHT technique

This section presents a review on various applications of signal processing techniques (used in this study) in different engineering fields such as Mechanical and Aerospace Engineering fields.

Babu et al. [449] found that the HHT was useful tool for extracting the salient features from time response of the cracked rotor passing through its critical speed. It has been observed that it is particularly useful for identifying very small crack depths where even CWT fails to detect them. The HHT could even give better results when used for a real system because the damaged system contains a component of nonlinearity. This is because it is based on the concept of instantaneous frequency rather than only resolution in frequency domain. Yan et al. [450] concluded that EMD-based HHT provided an effective tool for analyzing transient vibration signal as such it could effectively detect the deterioration of a test bearing through the extracted features including time-dependent amplitudes and instantaneous frequencies.

2.9.2 EMD technique

Cheraghi and Taheri [451] demonstrated the capability of energy-based damage indices extracted by EMD and FFT in identifying the existence and location of corrosion damages in pipes. Also, it was found that energy-based damage indices could also distinguish various levels of damages.

2.9.3 MUSIC technique

The application of MUSIC method for diagnosis of faults in sound signals of induction motors was reported by Garcia-Perez et al. [452] as shown in Figure 2.19. The validation of the proposed technique was done through the comparison against the analysis of vibration signal

From the application of MUSIC algorithm in damage detection of plate-type structures of aircrafts by Fan et al. [453], it was revealed that

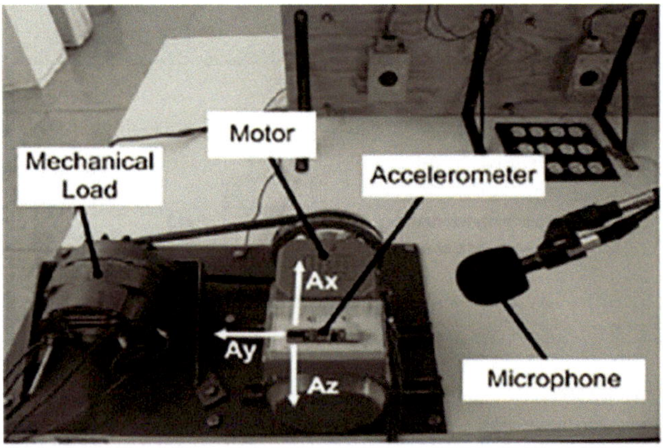

Figure 2.19 Test bench used during the experiment [452].

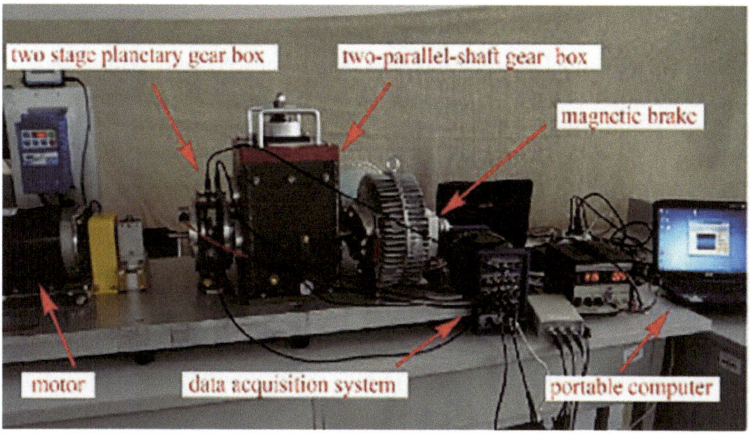

Figure 2.20 Fault experiment for planetary gear [455].

the MUSIC algorithm can not only effectively detect multiple damages in plate-type structures with good image quality but also has a super-resolution ability for detecting damage with distances smaller than half the wavelength.

2.9.4 CEEMDAN technique

The superiority of CEEMDAN in fault diagnosis of bearing signals in aircraft was revealed by Chi et al. [454] in comparison with singular value entropy, energy entropy, envelope sample entropy feature engineering and support vector machines classification model. In addition, from the application of CEEMDAN for diagnosing faults in planetary gear by Kuai et al. [455] as shown in Figure 2.20, the results demonstrated that CEEMDAN can accurately extract fault features generated from four different states of sun gears, and realize fault diagnosis of sun gears well because the IMFs decomposed by CEEMDAN contained the main characteristic information of planetary gear faults.

2.10 EXPERIMENTAL VERIFICATION OF CEEMDAN TECHNIQUE

In recent years, the HHT has been successfully used as a signal processing tool in SHM and damage investigation. The EMD and EEMD methods were proposed many times in previous studies as efficient tools for the purpose of damage identification. In order to investigate the capability and advantages of the CEEMDAN method as a successful technique in damage detection of a civil enginerring structure in comparison with the

EMD and EEMD, these three methods are utilized to analyze the measured structural response before and after the damage. Before applying this step, some advantages of CEEMDAN are explored and compared to the EMD and EEMD. Figure 2.21a–c show the comparison of the power spectral density (PSD) of the first three IMFs of acceleration response decomposed by the three methods. The reasons for the use of PSD in this study are: (i) to show at which frequency ranges, the variations of amplitude are strong, (ii) to indicate the predominant frequencies and neglecting the noisy spectrum lines generated by FFT. Since the PSD of each IMFs was not observed in the same frequency range, the logarithmic scale with base 2 is applied to the frequency range to properly present the comparison on a large-scale. In Figure 2.21c, it can be observed that the frequency spectra of each mode obtained by CEEMDAN are less overlapped, and obviously separated than those captured from EMD and EEMD in Figure 2.21a and b, respectively. This means that the CEEMDAN can solve the mode mixing problem of EMD and EEMD. In addition, the number of sifting iterations by CEEMDAN is about half of those from EEMD and EMD. Hence, CEEMDAN reduces computational time.

Figure 2.22 shows a one-second acceleration response of sensor 12 before the damage. The first three natural frequencies of the bridge are 19.53, 40.16, and 60.55 Hz, and the corresponding natural period times T_1, T_2, and T_3 are 0.051, 0.025, and 0.016 s, respectively. Hence, considering 1.0 s of the acceleration response of the bridge ($\approx 20T_1$) is reasonable to analyze

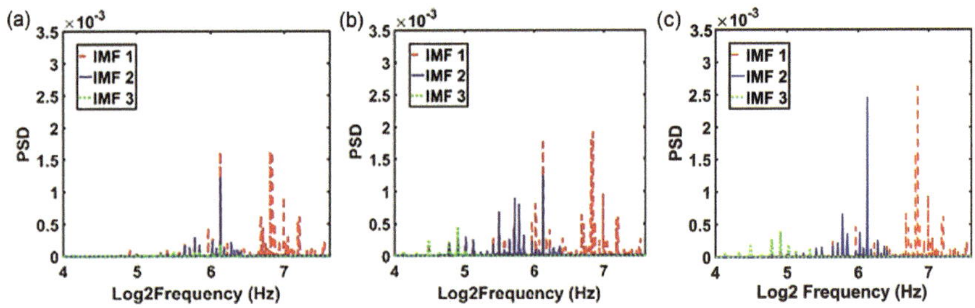

Figure 2.21 Spectra of IMFs 1 to 3 captured by; (a) EMD, (b) EEMD, (c) CEEMDAN.

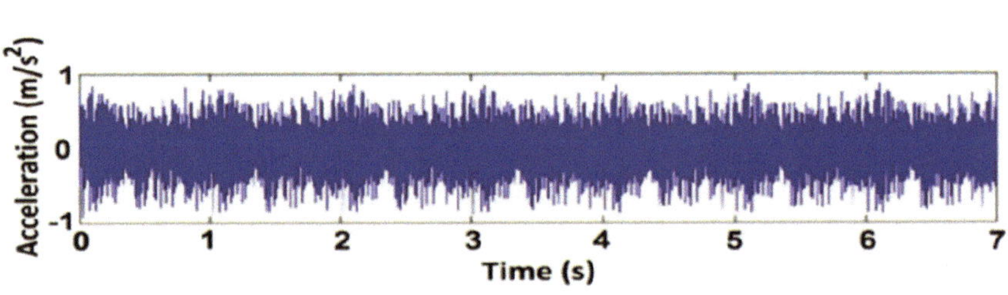

Figure 2.22 Acceleration response of sensor 12 in a healthy state of the bridge.

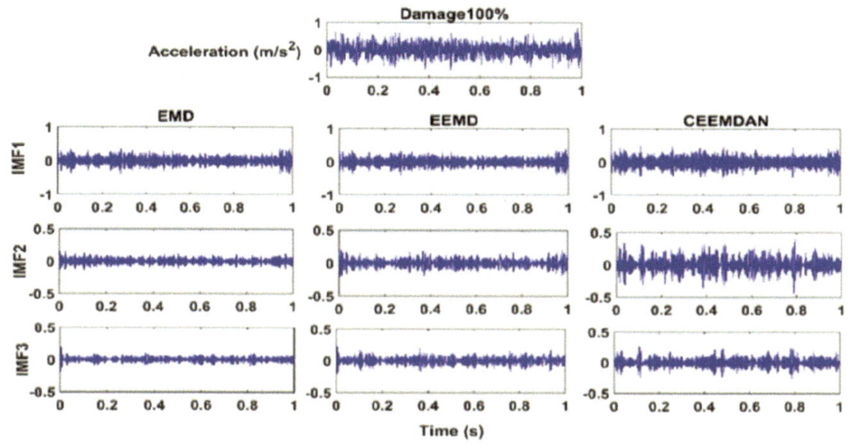

Figure 2.23 Comparison of IMFs from EMD, EEMD, and CEEMDAN for sensor 12.

in this study. In addition, the first three IMFs from an example acceleration response of sensor 12, decomposed by the three methods of EMD, EEMD, and CEEMDAN are compared in Figure 2.23. It is observed that CEEMDAN detects the spikes due to varying frequencies during the decomposition process more clearly compared to those from EMD and EEMD. Indeed, EMD and EEMD are not able to accurately reproduce the acceleration response of the truss. That is, the IMFs produced by CEEMDAN preserve the main characteristics and behaviors of the original signal such as the spikes and intensity of signal better than those obtained by EMD and EEMD. Note that CEEMDAN is the least affected by the mode mixing problem in decomposition. The following sections assess the application and performance of CEEMDAN compared to EEMD and EMD techniques in damage detection of the truss bridge.

In addition to the aforementioned advantages of CEEMDAN in solving the mode mixing problem, CEEMDAN also can preserve the original information of the analyzed signal. That is, the CEEMDAN presents a complete decomposition approach with an exact reconstruction of the original signal by summing the generated mode components as given in [30]. Due to this decomposition completeness of the CEEMDAN, this method can preserve the original information of the analyzed signal such as the intensity of the IMFs and existing sensitive spikes in the behaviors of the amplitude and energy features of the IMFs. In addition, the lower energy from the EMD and EEMD methods are also because of the effects of averaging over all realizations during the production of modes (that are independently produced from other realizations) while a large variation can exist in the number of modes. Accordingly, the CEEMDAN is significantly able to recover some information on the EMD properties lost by EEMD such as completeness and fully data-driven number of

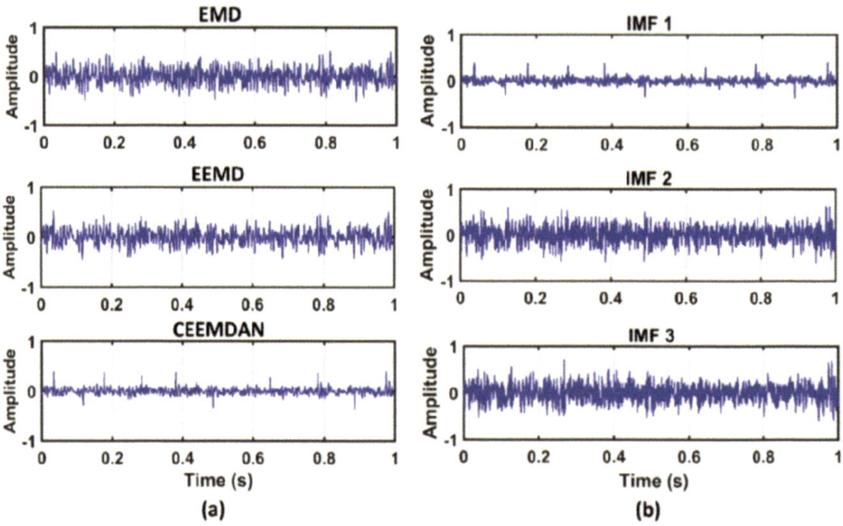

Figure 2.24 Reconstruction errors for; (a) CEEMDAN compared to those of EMD and EEMD, (b) IMFI compared to those of IMFs 2 and 3 decomposed using CEEMDAN.

modes. Therefore, it is expected that the results from the CEEMDAN to have more intensity than those from EMD and EEMD methods. In order to confirm this completeness and reconstruction effects of the CEEMDAN, Figure 2.24a shows the reconstruction errors computed as the difference between the acceleration signal recorded by sensor 12 and the sum of the modes using CEEMDAN in comparison with those from EEMD and EMD. It is seen that the maximum amplitude of the reconstruction error computed using CEEMDAN is very marginal compared to those from EMD and EEMD methods. Hence, the CEEMDAN represents a more sensitive approach compared to the previous generations of EMD technique. In addition, by comparing the reconstruction errors computed for the three first IMFs as the difference between the acceleration signal and each IMF in Figure 2.24b, it is found that the first IMF can more significantly reserve and reconstruct the information of the original signal compared to IMFs 2 and 3. Therefore, the intensity of the IMF1 and the corresponding features are larger than those derived from EMD and EEMD.

2.11 EXPERIMENTAL VERIFICATION OF CEEMDAN-MUSIC TECHNIQUE

Despite the advantages of CEEMDAN compared to previous generations of EMD-based techniques, It still have some drawbacks related to the remaining noise residues even after adding particular noise at each step of the decomposition. The existence of these noise residues may cause the generation of factious signal modes during the initial decompositions steps. To overcome this, the analysis of PSD of the IMF's with high-resolution spectral resulting from the application of MUSIC on the IMFs, can capture steady-state the analyzed signal.

In this section, the MUSIC and CEEMDAN-MUSIC techniques are verified using a synthetic signal as shown in Figure 2.25. This signal has been comprised of an initial 20.0 Hz sine wave, 25 Hz cosine wave at 0.2 s, two 30 Hz Ricker wavelets at 0.4 s, and 0.62 s, superposed 30 Hz Morlet atom at 0.5 s, and 70 Hz sine wave at 0.8 s. Figure 2.26a illustrates the MUSIC and CEEMDAN-MUSIC of the synthetic signal. In addition, the FFT of the original signal and its first IMF are shown in Figure 2.26b and c,

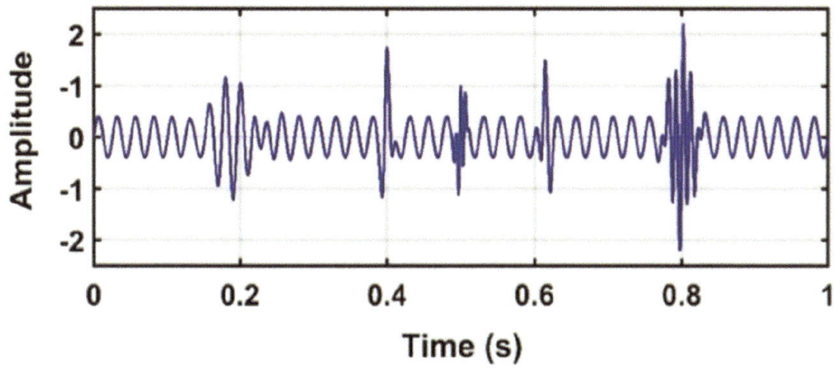

Figure 2.25 The synthetic signal contains the components of the background 20 Hz sine wave, 25 Hz cosine wave at 0.2 s, two 30 Hz Ricker wavelets at 0.4 and 0.62 s, superposed 30 Hz Morlet atom at 0.5 s, and 70 Hz sine wave at 0.8 s.

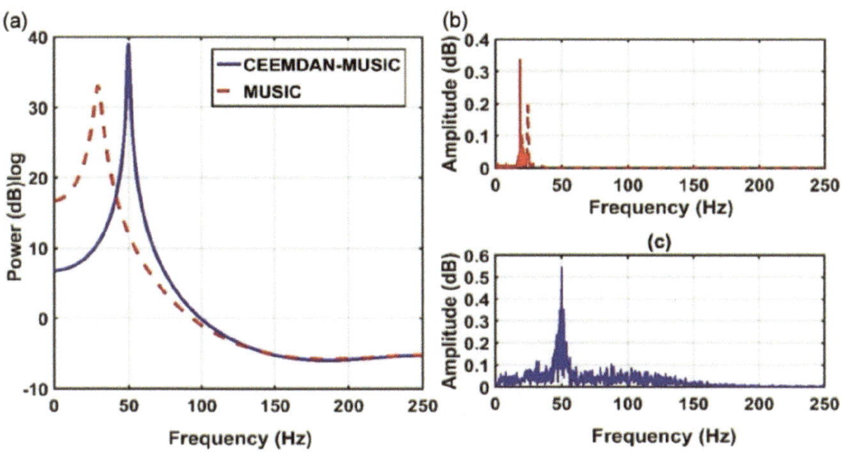

Figure 2.26 (a) Pseudospectra of the synthetic signal by MUSIC and CEEMDAN-MUSIC techniques, (b) the corresponding FFT of the original signal, (c) the corresponding FFT of the first IMF.

respectively. It is observed that the pseudospectral peaks of MUSIC and CEEMDAN-MUSIC occur at the same frequencies as those of the FFT peaks of the original signal and the first IMF around the frequencies of 45 Hz and 50 Hz, respectively. Therefore, the reliability of the proposed method to accurately estimate the pseudospectrum of non-stationary signals can be concluded. Therefore, the first IMF of CEEMDAN sufficiently represents the original information and contents of the analyzed signal (including modal information and natural frequency of the signal).

In the next Chapter, the application of HHT based on three decomposition method including empirical mode decomposition (EMD), ensemble empirical mode decomposition (EEMD), and complete ensemble empirical mode decomposition with adaptive noise (CEEMDAN) are investigated and compared in the damage detection process. Furthermore, the performance and of the sensitivity of different approaches such as multiple signal classification (MUSIC) and empirical wavelet transform (EWT) techniques in the damage detection and quantification. are assessed in this Chapter.

Chapter 3

Implementation of the proposed approaches in structural damage detection and quantification

3.1 INTRODUCTION

One of the most common techniques to analyze the nonlinear and non-stationary structural response is Hilbert-Huang Transform (HHT). Generally, conventional signal processing techniques such as Fast Fourier Transform (FFT) [109,382,456], and Short-Time Fourier Transform (STFT) [174,212] are employed to analyze the linear and stationary signals. However, the responses of structures in the real world are inherently nonlinear and non-stationary. Therefore, the need for adopting Hilbert-Huang Transform (HHT) [457,458] has been pronounced which is able to perfectly evaluate the nonlinear and non-stationary vibrations. The procedure of this technique contains two steps: (i) to decompose the original signal by one of the EMD-based techniques such as EMD, EEMD, and CEEMDAN into a series of complete and oscillatory components, named intrinsic mode functions (IMFs), (ii) to capture the instantaneous frequency and amplitude features using the application of Hilbert transform (HT) to the IMFs. Unlike the existing limited number of HHT applications based on EEMD and CEEMDAN, the vast majority of studies in the literature utilized HHT based on EMD for SHM and damage detection purposes [236,243,247,394,399,408,409]. The applications of HHT based on EMD, EEMD, and CEEMDAN techniques are evaluated in the following sections.

The uniqueness and special advantages of using the HHT based on CEEMDAN compared to conventional techniques such as FFT or wavelet is the feasibility of analyzing the nonlinear and nonstationary data and extracting the key features of the decomposed signal including IA, energy, unwrapped phase, and IF. These features of IMFs can reflect how the energy and phase of the signal vary with time. Moreover, the IF feature indicates the data in a time-frequency-power domain through the spectrogram plots. Besides, since the IMF is almost mono-component, all the instantaneous parameters from a nonlinear and non-stationary signal can be efficiently captured by HT. The information obtained by these features can provide a more accurate and real-life representation of the signal which can overcome

DOI: 10.1201/9781003499046-3

some shortcomings such as artifacts associated with the nonlocal and adaptive limitations generated by FFT and wavelet methodologies.

3.2 APPLICATION OF CEEMDAN TECHNIQUE

An efficient damage detection approach is introduced based on the application of the CEEMDAN method which has not been used in civil engineering. To evaluate the performance of the CEEMDAN-based damage detection approach, at first, the vibrations at the identified accelerometer sensors of the steel truss bridge model subjected to a band-limited white noise are acquired in the healthy and damaged states. Afterward, the vibrations are decomposed using the CEEMDAN technique to generate the set of IMFs. Then, the HT is applied to the first IMF to extract four features including energy, unwrapped phase, *IA*, and *IF*. Then, three different damage indices are defined based on the obtained vibration features from the healthy and damaged states including *IA*, energy, and unwrapped phase. In addition, four improved damaged indices based on the combinations of kurtosis and entropy features with each of the energy and *IA* features are proposed. Finally, the defined damage indices and HHT spectrums are adopted to identify the presence, location, and severity of the damage in the truss. The flowchart diagram of the proposed methodology is shown in Figure 3.1. Each step of this flowchart will be described in detail in the Results and Discussion section.

3.2.1 Detection of the presence and severity of damage using CEEMDAN technique

In this section, first, the existence and severity of the damage are studied by investigating the results from different damaged states (damage scenarios) of the bridge. To do this, three different damage levels are considered by reducing the cross-section stiffness (i.e., the moment of inertia). In this study, the damage states of the truss are implemented by reducing cross-section stiffness of a diagonal element and replacing it with three different damaged elements. The reductions of the cross-sectional moment of inertia of these damaged elements in percentage terms are; (i) 35% in which the outer and inner diameters are 16 and 10 mm, respectively as shown in Figure 3.2a, (ii) 60% in which the outer and inner diameters are 14 and 8 mm, respectively as shown in Figure 3.2b, (iii) 83% in which the diameter of the bar element is 11 mm as shown in Figure 3.2c. In addition, the axial stiffness reductions of these damaged elements are 14%, 27%, and 33%, respectively.

It is noteworthy that other structural parameters and loading conditions are kept constant for all the scenarios studied in this paper. In addition, the end conditions of the elements and are not changed by replacing the elements with different cross-sections. In this case, the elements are adjusted

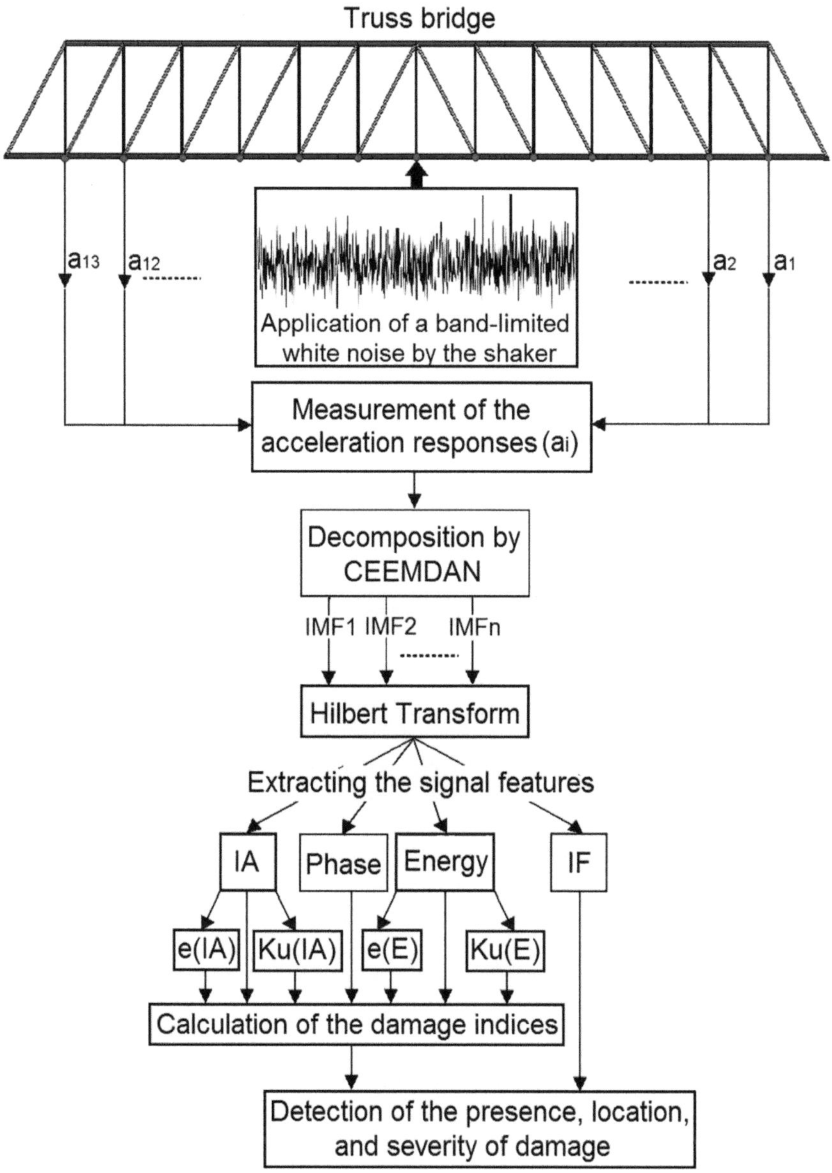

Figure 3.1 The framework of the proposed damage detection approach.

using identical screws and hinges (with identical sizes) that have been fixed to the joints. Therefore, replacing different elements does not cause the variations of the end conditions and consequently has no significant effects on the structural responses. To detect the presence and severity of the damage

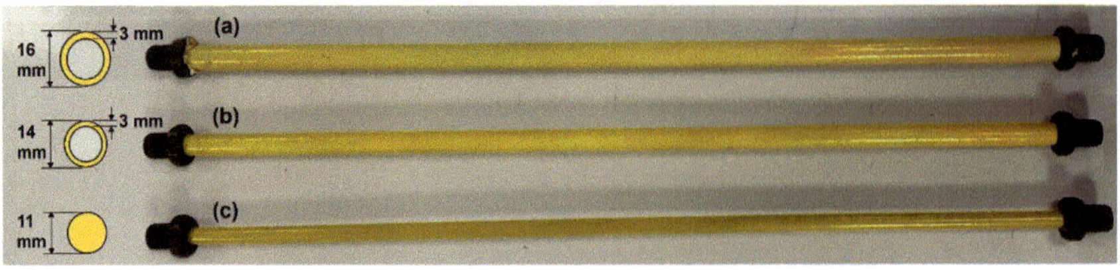

Figure 3.2 Details of the damaged elements with; (a) 35%, (b) 60%, and (c) 83% cross-sectional (i.e., the moment of inertia) stiffness reductions.

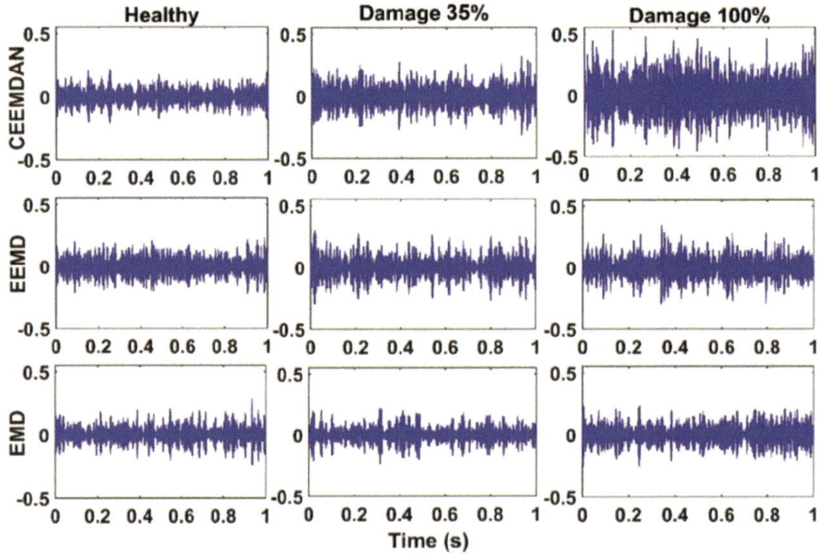

Figure 3.3 The first IMF results extracted by CEEMDAN, EEMD, and EMD techniques for sensor 10 for healthy and different damaged states of the structure.

in the truss, sensor 10 is selected as an example checkpoint located near the damaged element. Under these conditions, the first IMF extracted by the CEEMDAN, EEMD, and EMD techniques from the acceleration response of sensor 10 are presented before (i.e., healthy), and after damage states of the structure with levels of 35%, 60%, 83%, and 100%.

In Figure 3.3, the damage spikes are significantly seen in the time history behavior of IMFs for the damaged cases of the structure. By comparing the results of the first IMF extracted by CEEMDAN, EEMD, and EMD techniques, it is observed that the difference between the intensity of spikes of the IMFs becomes more pronounced with increasing the damage level, especially in the result from CEEMDAN. Furthermore, the values of the root-mean-square (RMS) which represent the total behavior of the IMFs,

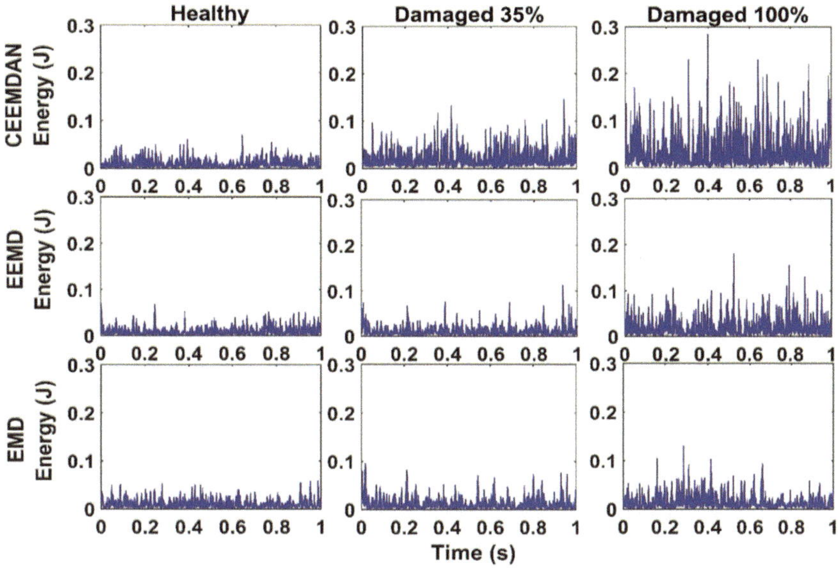

Figure 3.4 The first IMF's energy of sensor 10 was extracted by three techniques for healthy and different damaged states of the structure.

increase with increasing the damage level. Although these observations can show the superiority of CEEMDAN compared to EEMD and EMD techniques in the detection of damage, more accurate calculations based on the main features of the signal are required. Consequently, the Hilbert Transform is applied to the captured IMFs from the sensors, before and after damage, to extract four significant features including the energy, *IA*, unwrapped phase, and *IF*. Figure 3.4 shows the energy of the first IMF extracted from CEEMDAN, EEMD, and EMD techniques for the acceleration response of sensor 10, for healthy and different damaged states of the structure.

The increase of the energy is observed in proportion to the increase of the damage level. Moreover, it is obviously seen that the energy output extracted by CEEMDAN is more sensitive to the increase of the damage level compared to those of EMD and EEMD. Figure 3.5a illustrates the damage index results based on the energy feature of the first IMF by three EMD-based techniques for the acceleration responses recorded by all the sensors. It is seen that the energy-based damage index values calculated for all the damaged levels using CEEMDAN are about two times greater than those from EMD and EEMD techniques. Therefore, more sensitivity of the CEEMDAN technique in detecting the presence of damage in the vicinity of sensor 10 is concluded compared to other techniques.

In addition, the damage index results based on the combinations of the kurtosis and entropy features with the energy feature of the first IMF of

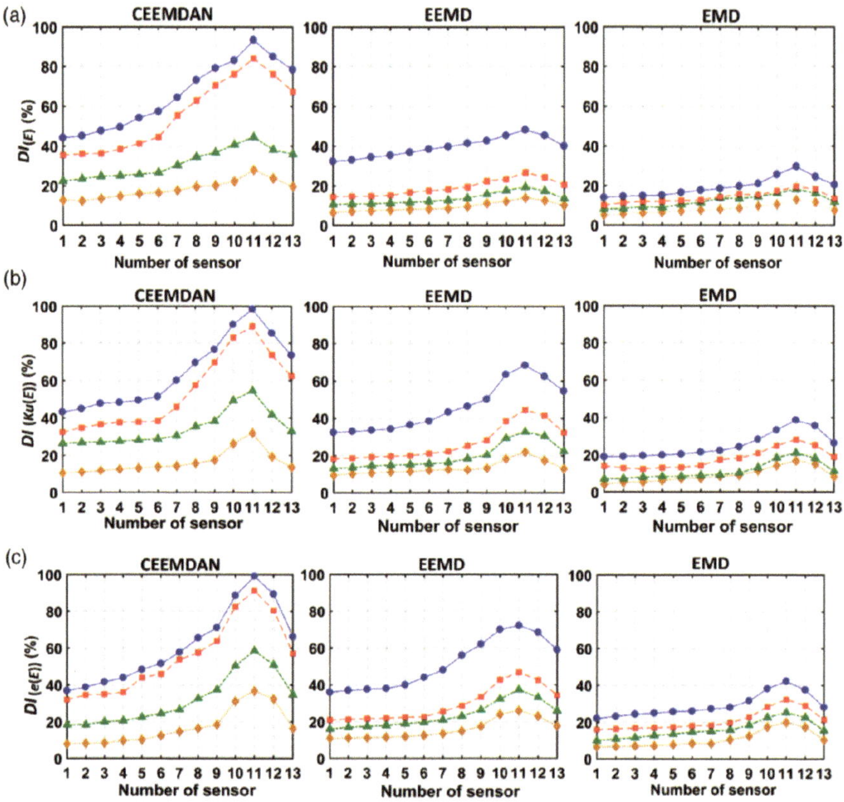

Figure 3.5 Comparing the damage index results computed based on the; (a) the energy of the first IMF ($DI_{(E)}$), (b) the kurtosis of the energy feature of the first IMF ($DI_{(Ku\ (E))}$), (c) the entropy of the energy feature of the first IMF ($DI_{(e\ (E))}$), using three techniques for the vibration responses of the bridge in different damaged states.

the acceleration responses recorded by all the sensors are illustrated in Figures 3.5b and c, respectively. It is seen that the trend of the curves of the improved damage indices (i.e., those computed from the statistical information of the energy of the first IMF) are more concentrated around the damage location (i.e., sensor 11) compared to those directly computed from the energy feature of the first IMF. Therefore, it is obtained that the CEEMDAN results not only capture higher damage index values compared to other techniques, but also the proposed damage indices are able to result in more sensitive trends in detecting the presence and location of the damage.

Moreover, the instantaneous amplitude (IA) feature is also assessed as another signal feature of the first IMF extracted by CEEMDAN, EEMD, and EMD techniques. In Figure 3.6, it is seen that the intensity of IA resulting from all three techniques increases with the increase of the damage

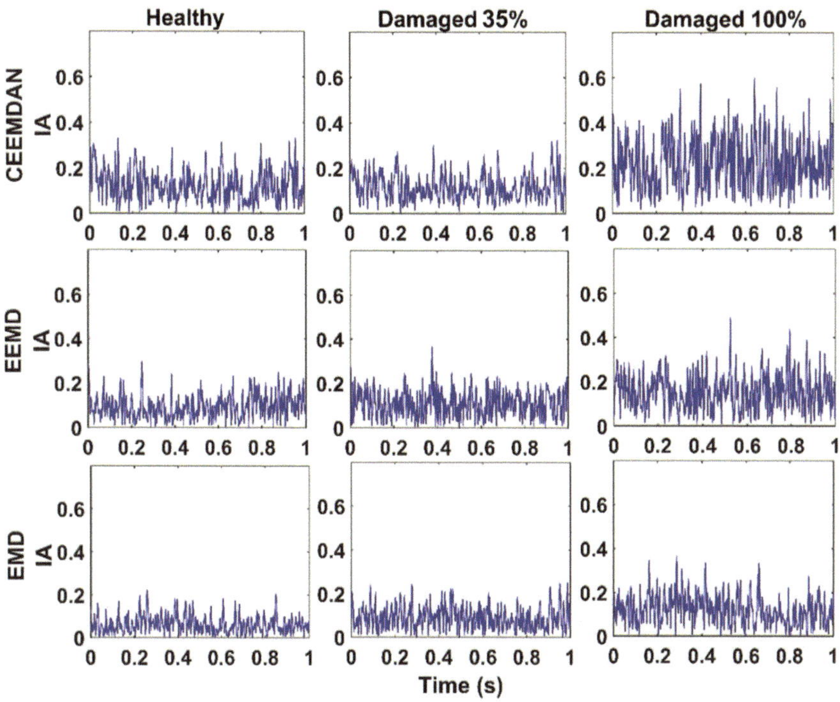

Figure 3.6 The Instantaneous amplitude of the first IMF of sensor 10 extracted by three techniques for healthy state and different levels of the damage.

level. However, the enhancement of *IA* resulting from CEEMDAN is more obvious compared to EEMD and EMD. To quantitatively investigate the sensitivity of the CEEMDAN to the presence and location of the damage in the truss in a more effective manner, the damage indices based on the *IA* feature of the first IMF of all the sensors from three techniques are computed and the results are illustrated in Figure 3.6a. It is seen that the damage index values from CEEMDAN are significantly higher (about two times) than those of EMD and EEMD.

In Figures 3.6b and c, the damage index results based on the kurtosis and entropy of the *IA* of the first IMF of all the sensors are illustrated. It is seen that the damage index curves have a similar trend to those observed in Figures 3.5b and c. This means that the CEEMDAN technique is not only more sensitive to the presence and location of the damage compared to EMD and EEMD but also its sensitivity becomes more pronounced when utilizing the improved damage indices based on the statistical information of the *IA* feature. The reason for the higher sensitivity of the proposed hybrid damage indices compared to those based on pure energy and *IA* features is related to the use of PDF expression by kurtosis and entropy features. This option causes the increase of the sensitivity of the indicator to any changes (even slight changes) in the analyzed vibration signal due to the stiffness change of the structure. Besides, as mentioned in Chapter 1 and proven in Chapter 4, the major advantages of the CEEMDAN are its completeness and reconstruction effects in the processing of a vibration signal that consequently preserves the dynamic information of the original signal such as peak values and sudden spikes. Hence, applying these statistical-based features to the energy and *IA* features extracted by CEEMDAN results in more sensitive damage index values compared to those of EMD and EEMD (Figure 3.7).

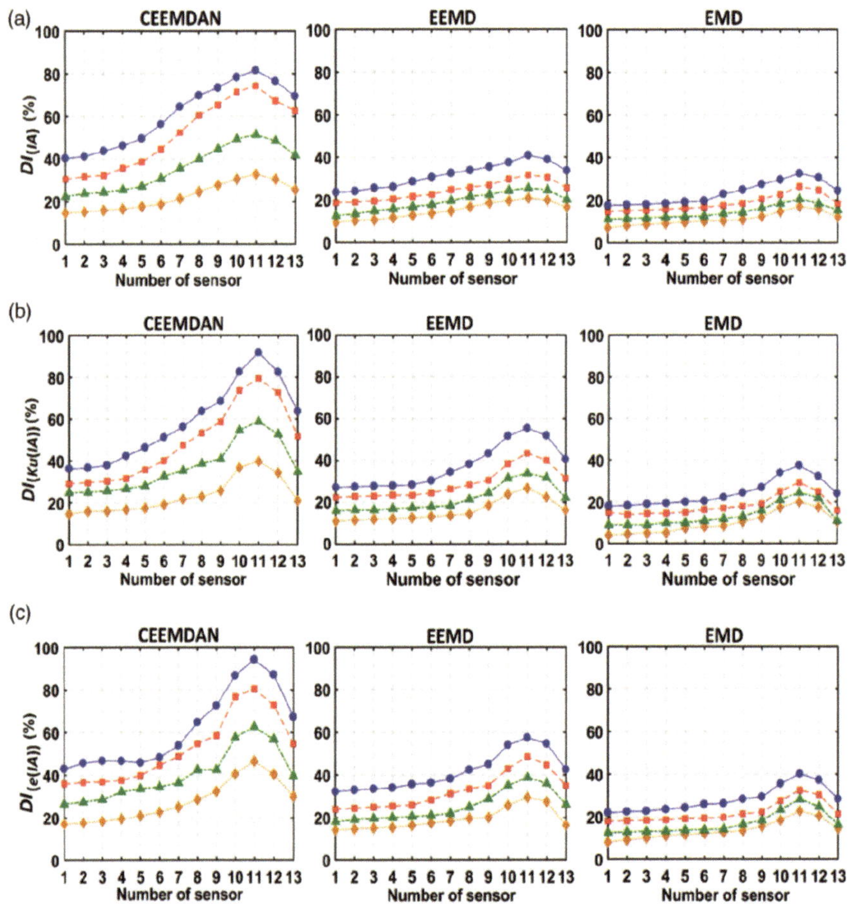

Figure 3.7 Comparing the damage index results computed based on (a) the instantaneous amplitude (*IA*) of the first IMF ($DI_{(IA)}$), (b) the kurtosis of the instantaneous amplitude of the first IMF ($DI_{(Ku\ (IA))}$), (c) the entropy of the instantaneous amplitude of the first IMF ($DI_{(e\ (IA))}$), using three techniques for the vibration responses of the bridge in different damaged states.

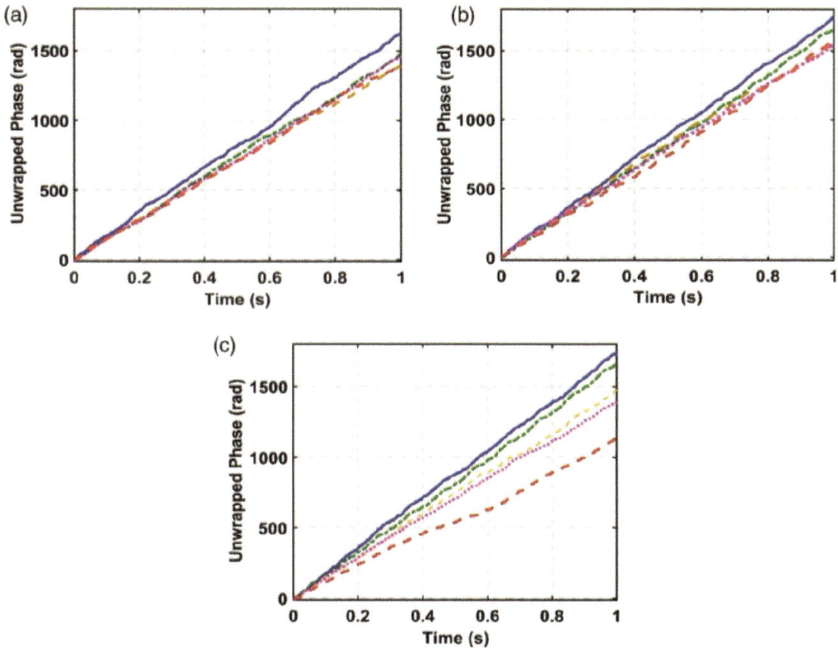

Figure 3.8 Comparison of the first IMF's unwrapped phase of sensor 10 by (a) EMD, (b) EEMD, and (c) CEEMDAN technique under four damage severity and healthy state of the bridge.

In addition, the effectiveness of adopting the CEEMDAN compared to EEMD and EMD in classifying the severity of damage using the unwrapped phase feature is assessed in Figure 3.8. It is obvious that the unwrapped phase of the first IMF decreased and their deviation relative to the healthy unwrapped phase increased as the level of damage increased. Moreover, by comparing the results illustrated in Figures 3.8a to c it is found that the CEEMDAN has substantially positive influences on detecting and classifying the damage severity compared to those from EEMD and EMD. Accordingly, a damage index is based on the deviation of damage states (35%, 60%, 83%, and 100%) relative to the healthy unwrapped phase. The results from the damage index based on the unwrapped phase of the first IMF of all sensors analyzed using three techniques are illustrated in Figure 3.9 for different damage states of the bridge. The results demonstrate the enhancement in the value of the damage indices by increasing the damage level using the CEEMDAN technique. However, it is obtained that EMD and EEMD are not accurately able to classify the levels of damage. Besides, although the EEMD succeeds in classifying the damage level, EMD fails in this assessment and cannot present an efficient performance in classifying the damage severities compared to the CEEMDAN.

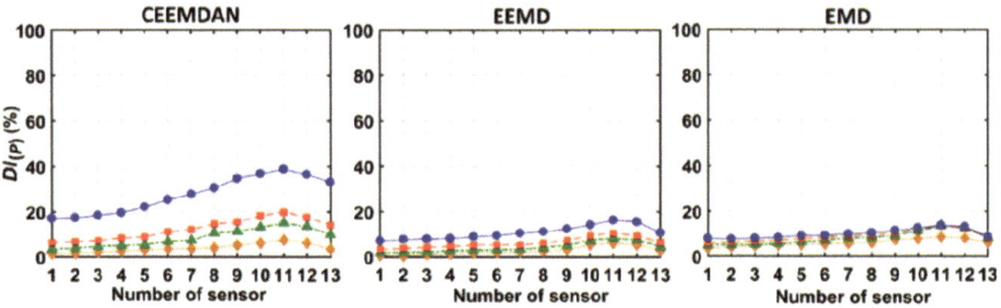

Figure 3.9 Comparing the damage index results among three techniques computed based on the unwrapped phase ($DI_{(P)}$) of the first IMF of all sensors for different damaged states of the bridge.

The instantaneous frequency is considered as another feature to detect the severity of damage through the HHT spectrum. This approach presents the instantaneous frequency of IMFs in a time-frequency domain with a sampling frequency of 500 Hz. The reason for adopting the HHT spectrum in this study is to demonstrate both instantaneous amplitude and instantaneous frequency features of the analyzed data in a time-frequency-energy domain with high resolution. There exist many research works in the literature utilizing HT spectrums to detect damage in linear systems [401]. The use of instantaneous frequency by HT spectrums has several key advantages including: (i) the recognition of frequency variations within one period, (ii) the identification of both inter-wave and intra-wave frequency modulations in a wave train which cannot be obtained using Fourier spectral analysis (i.e., Fourier spectral analysis can only capture the inter-wave frequency modulation), (iii) the ability to simulate the nonlinear and non-stationary data

According to the literature [401,436,459,460], it has been proven that the energy of HHT spectrums and the vibration magnitude increase at lower frequency ranges with the increase of damage level due to the stiffness loss of the structure. In other words, the damaged structure exhibits different degrees of frequency lowering which represents the presence of structural nonlinearity in the system caused by stiffness loss. This frequency reduction behavior is due to the decrease of the square root of the stiffness of the structure in the presence of the damage [460].

Figures 3.10a–c illustrate the energy spectrum of the IMFs in the time-frequency domains for the bridge in healthy and different damaged states. It is seen that the energy density of IMFs increases in the spectra in approximately the range of 0–20 Hz, with increasing damage level. In other words, the frequency has a high power between the ranges 25 and 30 Hz for the healthy state of the truss (see Figure 3.10a), as the level of damage

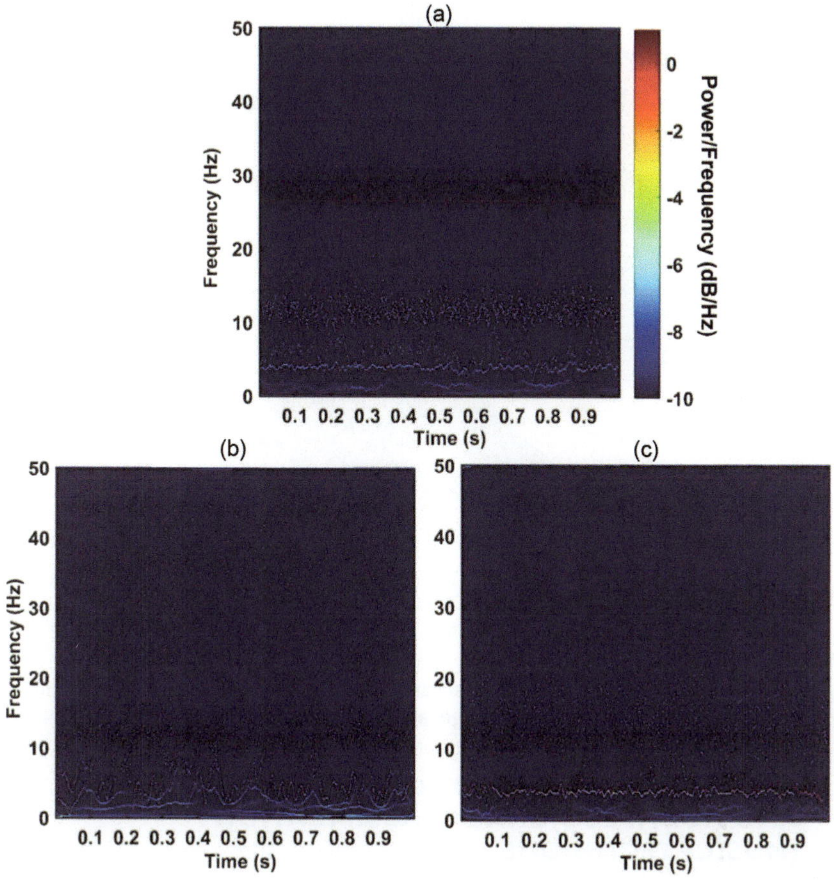

Figure 3.10 HHT spectrum of the sensor 10 by CEEMDAN for different states of the truss including; (a) healthy, (b) 35% damage, and (c) 100% damage.

increases, the energy density of normalized frequency increases in lower ranges (between the ranges 5 and 15 Hz) especially for 100% damage level as observed in Figure 3.10c. In addition, the three-dimensional spectrograms of the IMFs are presented in Figures 3.11a to c to show the HHT spectrums with high resolution in which the amplitude of the signal is given in a time-frequency domain with a sampling frequency of 500 Hz and the overlapping windows are 100 samples in length. In Figures 3.11b and c, it is seen that the intensity (power) of *IF* increases in the low-frequency range when the level of damage increases.

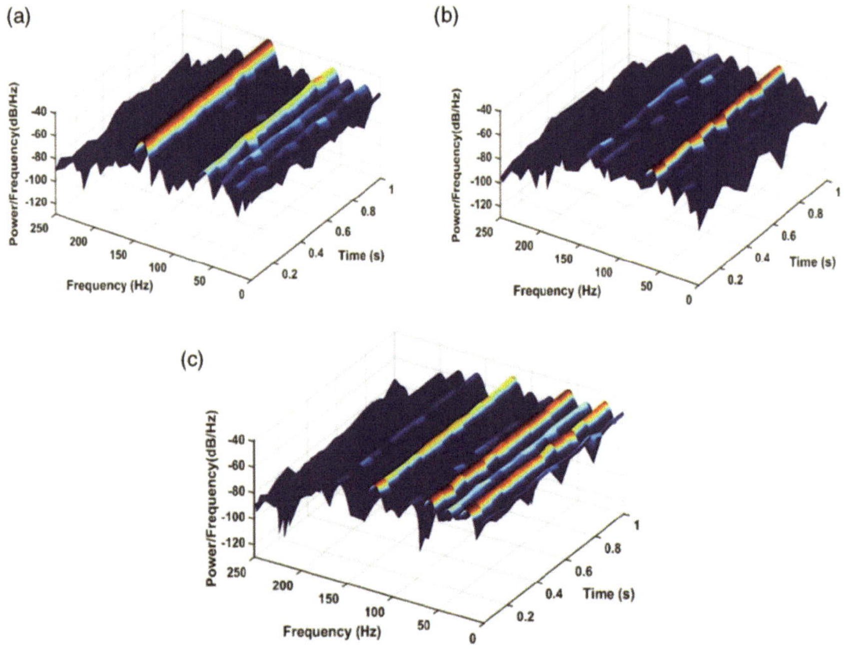

Figure 3.11 Spectrograms of the IMFs of sensor 10 for three states of the bridge including; (a) healthy, (b) 35% damage, and (c) 100% damage.

3.3 HYBRID APPLICATION OF CEENDAN AND MULTIPLE SIGNAL CLASSIFICATION (MUSIC) TECHNIQUE

The MUSIC algorithm provides a high-resolution power density spectrum which effectively improves the detectability of close frequencies in comparison with the performances of traditional techniques such as FFT in analyzing signals with low signal-to-noise ratios (SNRs) [173]. Therefore, as an extension of the previous research works, a hybrid damage detection approach is proposed in this study from the combined use of the MUSIC algorithm with CEEMDAN for damage detection of a steel truss bridge. Jiang and Adeli [461] proposed a new approach on the basis of the MUSIC algorithm for a scaled model of a 38-story RC structure. The results of the output predicted by the trained neural network model and the output measured by the sensors were compared and the difference between them was considered as an effective indicator of damage. Rios and Sanchez [178] used a fusion of MUSIC-ANN algorithms to address the location of damage and quantify the damage severity of a five-bay truss structure. To do that, the amplitude and the natural frequencies of the structural responses were used as the input data in an ANN. The result showed that the proposed method was effectively simple to use, automated, and reliable tool in SHM. MUSIC algorithm has also been used in mechanical structures. For instance, Perez et al. [462] proposed a fault detection procedure based on MUSIC for the purpose of high-resolution spectral analysis to enhance the anomaly detectability in an induction motor. Martinez et al. [463] combined EMD and MUSIC techniques to detect the mixed faults in induction motors. The results showed the advantage of the proposed approach which could identify the characteristic frequencies of the multiple combined faults.

Although the combination of MUSIC with EMD technique can be found in the literature, EMD has several drawbacks as mentioned above that can be solved by using CEEMDAN. Hence, the main motivation of this paper is to experimentally assess the performance of a combined CEEMDAN-MUSIC approach compared to pure MUSIC and several frequency-domain techniques such as FFT, power spectral density (PSD), and FDD in identifying the location and quantifying the damage severity of a scaled truss bridge model. To this end, the CEEMDAN is firstly used to decompose the acceleration responses of a laboratory-scale model of a steel truss bridge when exposed to a band-limited white noise. Then, the power density spectrum (named pseudospectrum) of the extracted IMFs is computed by the MUSIC technique prior to and post the damage states of the truss.

3.3.1 Procedure of the CEEMDAN-MUSIC damage detection method

To investigate the performance of the proposed combined CEEMDAN-MUSIC technique in damage detection of the truss bridge, first, the vibrations from the identified accelerometer sensors are acquired when healthy and damaged states of the steel truss bridge model are exposed to a band-limited white noise. Then, the vibrations are decomposed by the CEEMDAN method to produce a set of IMFs. Afterwards, the MUSIC algorithm is applied to the IMFs to transform the time series acceleration response to the power density spectrum domain (named pseudospectrum). It is noteworthy that in this study, the MUSIC algorithm is applied to only IMF1. Finally, the absolute differences between the pseudospectral peaks in terms of the frequency (D_f) (normalized using the pseudospectral peak from the healthy case) and the power magnitude (D_p) ranges are considered as the damage indicators. By evaluating these damage indicators, the presence, severity, and location of damage are determined. The performance of the proposed combined CEEMDAN-MUSIC technique is assessed by comparing the damage indicator results with those from the pure MUSIC algorithm. Figure 3.12 illustrates the framework flowchart of the proposed approach.

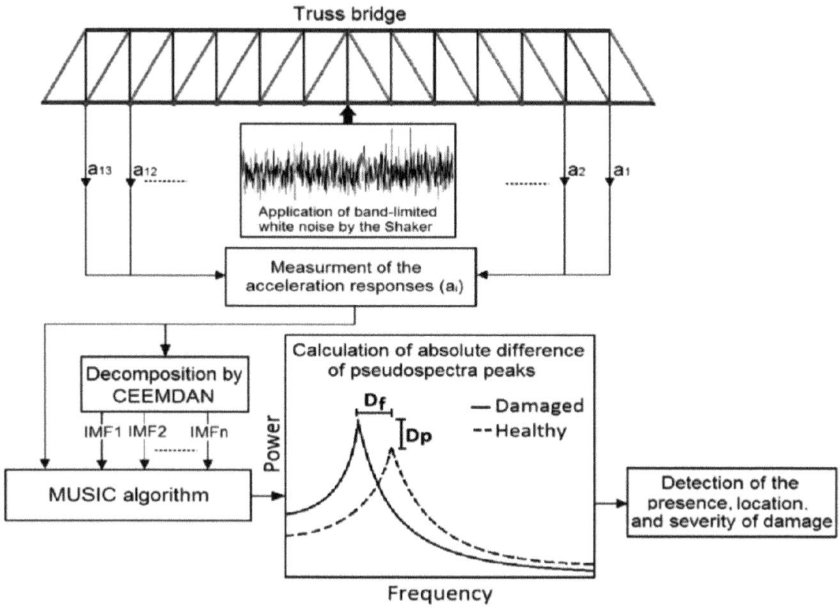

Figure 3.12 The framework of the proposed methodology (a_i denotes the acceleration responses)

Although CEEMDAN significantly reduces the computational cost by decreasing the number of useless IMFs generated in EMD and EEMD algorithms, it has a limitation in selecting the signal-noise-ratio (SNR) which is extremely dependent on the characteristics of the analyzed signal. As such, during the decomposition process of a signal by CEEMDAN, notable overlapping may be generated between the modes due to adding a particular EMD mode of white noise with different signal noise ratios (SNR) at each step.

Therefore, an empirical approach is needed to select an efficient coefficient of SNR to analyze a particular signal. Besides, the implementation process of the proposed methodology in this study which assesses a scaled model of a truss bridge built in a laboratory environment has several limitations compared to those employed in real environment listed as follows:

- A preprocessing step to denoise the signals of real-world structures in noisy environments should be performed.
- The effect of various factors on the vibrational measurements such as the evolution of boundary condition and ambient condition (e.g., temperature, traffic, etc.) should be considered for the evaluation of real-world structures.
- Since the case study is a laboratory-scale structure with low-dimensional elements compared to full-scale structures in the real world, the responses of the truss are acquired from the sensors attached to the joints

of the lower chord of the bridge with a uniform distribution. Although the joints are common locations for sensor attachment, an exact determination is more convenient and possible for full-scale bridge structures by attaching the sensors on the elements with higher dimensions.

- Due to manufacturing limitations of steel truss elements, the damage scenarios were performed by using only the damaged elements with producible diameters in the factory (only tubes with a thickness of 3 mm) as considered in this study with percentage reductions of the cross-section area of 35%, 60%, and 83%.
- To avoid redundancy and due to page limitation of this chapter, the damage scenarios of the truss were implemented by assuming the damage states for only one diagonal truss element.
- Since this paper investigates a laboratory-scale truss bridge under controlled conditions, it is assumed that the truss is subjected to only white noise excitation and the effects noise intrinsic to the system from unidentified sources are not considered. White noise is a random signal having equal intensity at different frequencies which can be resulted from various environmental excitations such as traffic, wind, walking in the real world.
- Artificial intelligence methods are needed to be utilized to classify the huge amounts of data generated by large-scale and complex structures in the real world.

3.3.2 Detection of the presence and severity of damage using CEEMAN-MUSIC

The existence of the damage in the truss and its severity is studied in this section by analyzing the experimental responses of the bridge with different damage states (i.e., damage scenarios). Figure 3.13a shows a 7-second segment of the acceleration response of the truss recorded by sensor 10 prior to applying damage. In addition, the corresponding FFT, and FFT of 100% damage state of the truss are illustrated in Figures 3.13b and c, respectively. The first three natural frequencies of the bridge in a healthy state are 19.53, 40.16, and 60.55 Hz, and the corresponding natural period times T_1, T_2, and T_3 are 0.051 s, 0.025 s, and 0.016 s, respectively. Therefore, selecting one second of the acceleration response (about $20T_1$) seems a sufficient time window to be analyzed. In Figure 3.13c, although the natural frequencies are slightly reduced by applying the damage in the truss, the FFT method has some difficulties since this technique is not able to clearly address the presence of the damage and classify its severity due to existing many spurious peaks. As mentioned in Section 3.2, the motivation of this paper is to propose a new damage detection approach based on the combination of CEEMDAN and MUSIC techniques to clearly detect the damage and classify its severity in comparison with the performances of several conventional frequency domain methods such as standard FFT, PSD, FDD, and pure MUSIC in damage detection. First, one second of the acceleration response

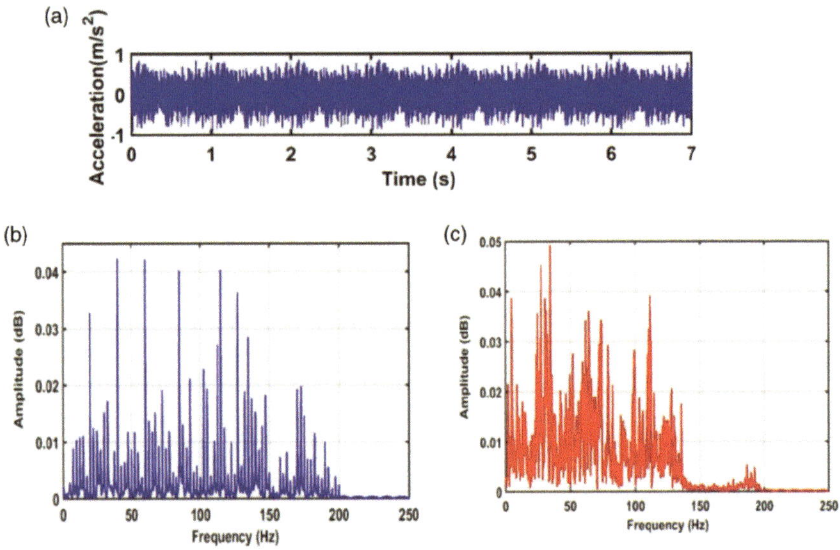

Figure 3.13 (a) Acceleration response of the bridge recorded by sensor 10 in a healthy state, (b) corresponding FFT, (c) FFT of the bridge response for 100% damage state of the truss.

of sensor 10 is decomposed by CEEMDAN before applying the damage to the truss as shown in Figure 3.14. The first three IMFs extracted by the CEEMDAN technique from the acceleration response recorded by sensor 10 attached to the healthy and damaged states of the bridge with damage severities of 35%, 60%, 83%, and 100% reduction in the cross-section stiffness are respectively presented in Figure 3.15 from left to right.

In Figure 3.15, the sudden spikes observed in the time history of IMFs represent the presence of damage in the structure similar to observations by the previous work [283]. Moreover, the intensities of these spikes and the RMS values representing a total trend of IMFs significantly increase with the increase of damage level. Besides, it is seen that the intensities of signal spikes and the RMS values of the first IMF are significantly higher than those of second and third IMFs. Therefore, since the first IMF gives the most energy value and contains the highest frequency content of the measured signal (RMS amplitude), it is appropriate to be considered as the index IMF in the damage detection procedure of this paper. In addition, based on an investigation carried out in the previous study [212], the capability and advantages of the CEEMDAN in solving the mode mixing problem associated with EMD and EEMD techniques, and preserving the original information of the analyzed signal were concluded. Hence, the CEEMDAN can present a complete decomposition approach with an exact reconstruction of the original signal by summing the generated mode components. Therefore, adopting the CEEMDAN technique can effectively preserve the original

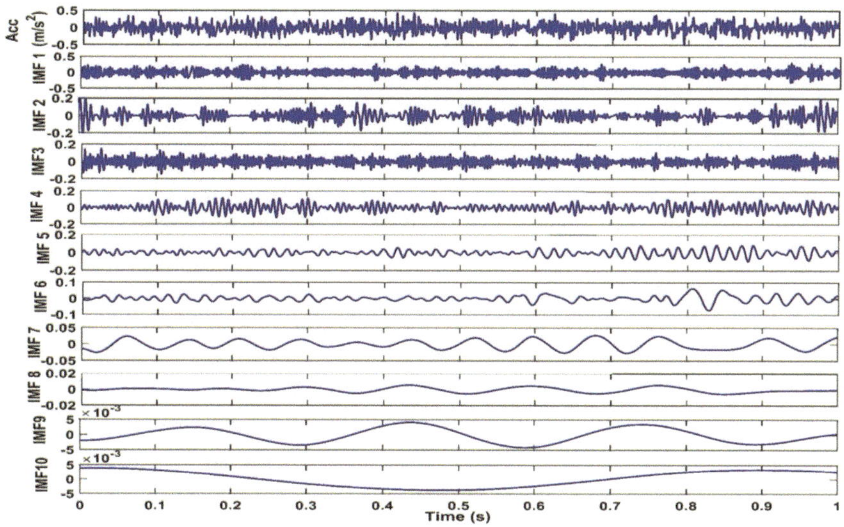

Figure 3.14 Decomposition of the acceleration from sensor 10 by CEEMDAN prior to applying damage with a noise standard deviation of 0.2 and an ensemble size of $I = 100$ (Acc denotes Acceleration).

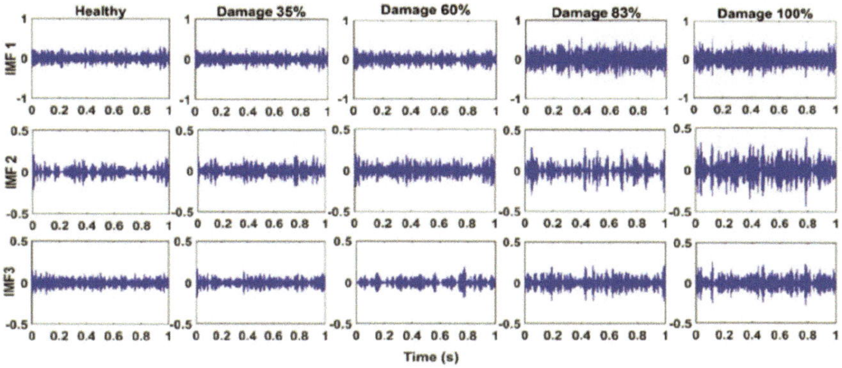

Figure 3.15 Time histories of the first three IMFs by CEEMDAN for the acceleration response of sensor 10 recorded for healthy and damaged states of the bridge.

information of the analyzed signal, such as the intensity of the IMFs and existing sensitive spikes in the behaviors of the amplitude and energy features of the IMFs. More detailed information about the superiority assessment of the CEEMDAN compared to traditional EMD-based techniques can be found in the previous study. [212]. The performance of several traditional spectral analysis methods including FFT, PSD, and FDD techniques in identifying the presence and severity of damage in the truss is firstly evaluated. To this end, the acceleration responses of the bridge in health and damage states recorded by sensor 10 (an example checkpoint) are analyzed. Thereafter, the FFT, PSD, and FDD techniques are employed to analyze the truss vibration recorded by sensor 10 for healthy and damaged states with levels of 35%, 60%, 83%, and 100% and the results are illustrated in Figures 3.16a–c, respectively. It is observed that FFT, PSD, and FDD fail in accurately characterizing the severity of the damage. They can only demonstrate the suspected presence of the damage by the formation of a spike in the low-frequency ranges. Therefore, the FFT, PSD, and FDD algorithms are not efficiently able to classify the damage level. In comparison, from Figure 3.17a, it is seen that the pseudospectral results from the pure MUSIC

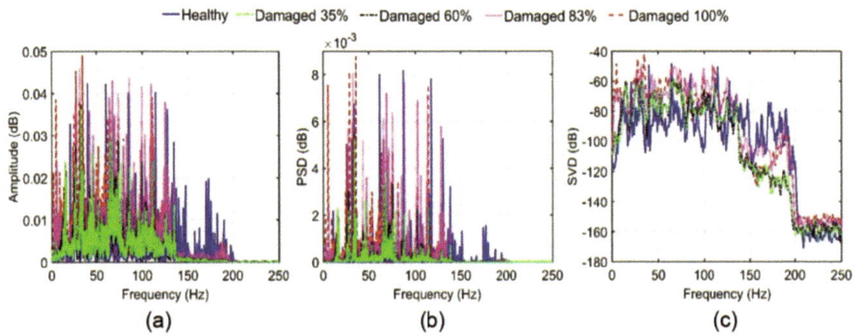

Figure 3.16 Comparing; (a) FFT, (b) PSD, and (c) SVD of the acceleration responses of the truss for healthy state and different damage levels of 35%, 60%, 83%, and 100%.

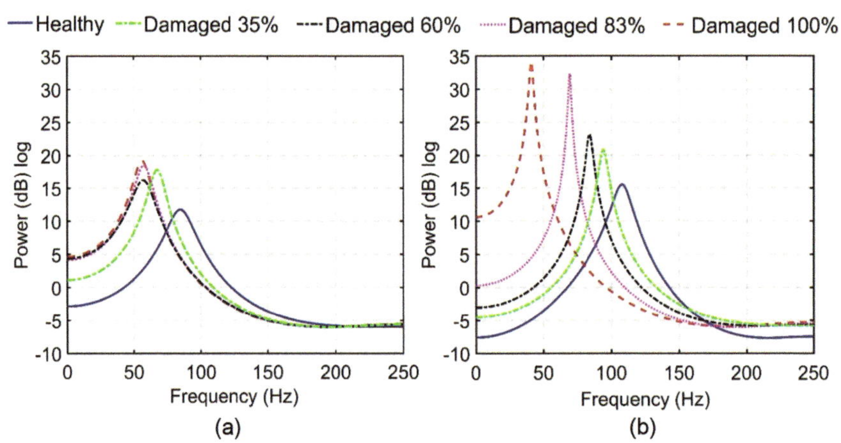

Figure 3.17 (a) Comparing the pseudospectrum of the acceleration responses of the truss obtained by pure MUSIC algorithm, (b) comparing the pseudospectrum of the IMF1 of the acceleration responses by CEEMDAN-MUSIC technique for healthy state and different damage states with levels of 35%, 60%, 83%, and 100%.

technique clearly can demonstrate the existence of damage in low-range frequencies compared to the frequencies of the bridge in a healthy state. In this case, the differences between the pseudospectral peaks of the healthy and damaged states are considered as the damage indicators. While the damage levels are not still distinguishable. To overcome this drawback, the MUSIC technique is combined with CEEMDAN, named CEEMDAN-MUSIC in this study. The proposed methodology represents the application of MUSIC on the first IMF extracted by CEEMDAN, instead of directly performing it on the acceleration response.

As shown in Figure 3.15, it was found that CEEMDAN properly demonstrates the increase of damage levels, through the increased intensity of peaks in the IMFs. Hence, CEEMDAN is utilized to improve the performance of the MUSIC technique. Also, the first IMF is selected for use in combination with the MUSIC technique because it contains the most energy of the original acceleration response of the bridge. Figure 3.17b shows the pseudospectral results from CEEMDAN-MUSIC. The significant differences between both frequencies and pseudospectral power magnitude of the peaks are observed. That is, as the level of damage increased the distance among the peaks in the five pseudospectral substantially. Thus, the proposed technique can efficiently detect the presence and classify the damage level. Furthermore, Figure 3.17b demonstrates the improvement of damage detection and classification by CEEMDAN-MUSIC compared to pure MUSIC. The effectiveness of the CEEMDAN technique in combination with MUSIC can be controlled by several key parameters. From a series of convergence tests based on the trial-and-error approach, the results converged when noise standard deviation (Nstd), number of realizations (NR), and maximum number of sifting iterations (MaxIter) were 0.2, 100, and 1000, respectively.

In addition, Table 3.1 presents the values of absolute differences of pseudospectral peaks in frequency (D_f) (normalized using the peak of the undamaged case) and power magnitude (D_p) ranges, as a powerful indicator that can efficiently detect the presence of damage and its severity. In addition, it is clearly observed that indicator values of CEEMDAN-MUSIC are more than two times greater than those from the MUSIC alone.

Table 3.1 The absolute difference of spectrum peaks in the ranges of the normalized frequency (D_f) and the power magnitude (D_p) before and after four different damage levels

Damage severity (%)	*MUSIC*		*CEEMDAN-MUSIC*	
	D_f (Hz)	*D_p (dB)*	*D_f (Hz)*	*D_p (dB)*
35	0.21	6.03	0.12	5.26
60	0.32	4.43	0.21	7.35
83	0.32	6.75	0.35	16.78
100	0.34	7.42	0.62	18.65

3.4 CHAPTER SUMMARY

In the first part of this chapter, the performance of the EMD-based signal processing technique referred to as CEEMDAN was experimentally assessed in identifying the presence and severity of the damage for a steel truss bridge model. Based on the evaluations using three signal processing techniques including EMD, EEMD, and CEEMDAN, the first IMF extracted by the CEEMDAN was selected as the evaluation criterion in detecting the structural damage. In addition, four key parameters of the signal including the energy, instantaneous amplitude (*IA*), unwrapped phase, and instantaneous frequency (*IF*) extracted through applying Hilbert transform (HT) to the IMFs are considered to investigate the existence and severity of the damage in the model. Furthermore, the sensitivity of CEEMDAN compared to the previous generations of the EMD-based techniques was investigated by proposing several improved damage indices from the combinations of two statistical signal features including kurtosis and entropy with the energy and instantaneous amplitude features of the analyzed signal.

The main findings and contribution of the first part of this chapter can be summarized as follows:

- The CEEMDAN approach as a novel extension of EMD accurately reproduces a proceed signal and mitigates the mode mixing problem.
- The intensity of spikes of the first IMF decomposed by CEEMDAN increased more significantly compared to those from EEMD and EMD techniques, which demonstrates the completeness of CEEMDAN in decomposing the acceleration response of the structure. Therefore, the CEEMDAN presented a more sensitive damage detection approach compared to EMD and EEMD.
- Increasing the energy and *IA* values and decreasing the unwrapped phase values of the first IMF were observed with the appearance of the damage, with increasing the level of damage, and with decreasing the distance of the desired sensor from the damage location.
- Although EMD and EEMD were able to detect the damage, a significant improvement of CEEMDAN compared to the other techniques in detecting the existence and severity of the damage using the damage indices based on energy, *IA*, and unwrapped phase parameters, was concluded.
- By assessing the *IF* of IMFs in the time-frequency-energy domain, an increase in the power of the frequency in the low ranges is observed when the damage appears. Furthermore, this technique can provide an indication of the general damage region, as well as being a sensitive indicator of damage.
- By comparing the value of the damage indices based on energy, *IA*, and unwrapped phase features, it was found that the energy feature represents a better approach to detecting, and classifying the severity of the damage.

- It was concluded that the proposed damage indices based on the combinations of the kurtosis and entropy features with the energy and *IA* features resulted in more sensitive indices in identifying the presence, intensity, and location of the damage compared to those based on the direct utilizing of the energy and *IA* features. In addition, more sensitive indices were obtained when combining the entropy feature with the energy and *IA* compared to those in which the kurtosis feature was utilized.

The second part of this chapter experimentally assessed the performance of a hybrid CEEMDAN-MUSIC technique as a novel signal processing technique in identifying the presence and classifying the level of damage in a laboratory-scale model of a steel truss bridge. The implementation process of this technique begins by employing the CEEMDAN algorithm to decompose the acceleration response of the truss bridge exposed to a white noise excitation. Then, the MUSIC technique was applied to the first IMF before and after damage to compute the power density pseudospectrum. In addition, the absolute differences of the spectral peaks in the frequency (D_f) and power magnitude (D_p) ranges extracted from the responses of different damage scenarios varying in terms of the damage severity are considered as the effective indicators to classify the level and detect the location of the damage. Significant improvements were observed in the damage detection and characterization process of the truss using the proposed method by comparing the results from the indicator values of the combined CEEMDAN-MUSIC algorithm with those from pure MUSIC technique and several frequency-domain techniques such as FFT, PSD, and FDD. As such, compared to many spurious peaks captured by the conventional frequency domain techniques leading to notable difficulties in the damage detection process, the proposed method was clearly able to address the classification of the damage severity levels and the damage location, as well as the perception of damages at lower levels. The contribution of this chapter is the robustness of CEEMDAN, MUSIC, and their combination in damage detection of civil engineering structure for the first time. As a result, the method can be recommended for health monitoring of various structures according to the proposed framework in this paper.

In Chapter 4, the experimental verification of the proposed damage detection using EMD-based techniques and their comparison, HHT, and MUSIC techniques are investigated in damage localization of the truss bridge model.

Chapter 4

Implementation of the proposed approaches in structural damage localization

4.1 DETECTION OF DAMAGE LOCATION USING CEEMDAN TECHNIQUE

In this section, the merit of the CEEMDAN-based damage detection approach in identifying the location of damage in the truss is assessed. To do this, the vibration responses of the structure in the healthy and damaged states were acquired from different four sensors of 11, 9, 4, and 1 which are the nearest, the second near, far, and the farthest sensors relative to the damaged element, respectively are analyzed. It is noteworthy that only a damage scenario with a 100% damage scenario is considered in this section by removing a diagonal element from the truss. Thereafter, the vibrations collected from the aforementioned sensors are decomposed by CEEMDAN, EEMD, and EMD techniques as shown in Figure 4.1.

In this study, it is noteworthy that the responses of the truss are acquired from the sensors attached to the joints of the lower chord of the bridge with a uniform distribution, not directly from the elements. Due to the existing limitations for the arrangement of the sensors on the elements of the laboratory-scaled truss bridge (because of lower-dimensional elements), a uniform distribution was selected in this study. Although the joints are also common locations for the attachment of sensors, an exact determination is more convenient and possible for full-scale bridge structures by attaching the sensors to the elements with higher dimensions.

By comparing the first IMF from the three techniques in Figure 4.1, it is observed that the intensity of spikes and the RMS value of the first IMF for closer sensors to the damaged element are greater than those of sensors located at farther distances. Furthermore, the energy of IMFs increases with decreasing the distance of the sensors from the location of the damaged element as shown in Figure 4.2. These higher spikes in the behaviors of the first IMF and its energy around the location of the damage (at sensor 11) lead to larger damage index values (about two times) compared to those from other techniques which can be clearly seen in Figure 3.5. The theoretical reasons for observing these notable spikes and enhancement trends in

 DOI: 10.1201/9781003499046-4

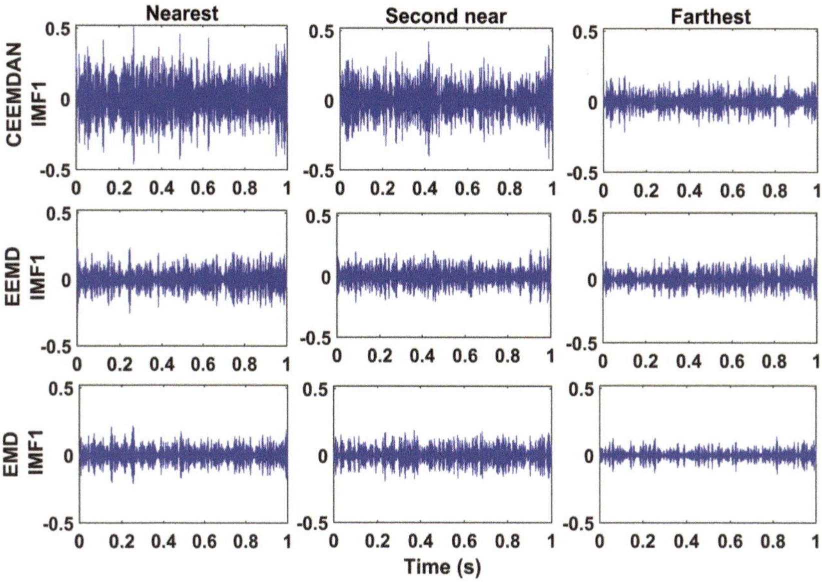

Figure 4.1 The first IMF results extracted by CEEMDAN, EEMD, and EMD techniques for different sensor locations.

the behaviors of the first IMF and its energy feature for closer sensors to the damage location resulting from CEEMDAN is due to the completeness of this technique in decomposing the original vibration signal with an exact reconstruction compared to EMD and EEMD. As discussed in Chapter 3, the CEEMDAN is able to preserve the key information of the original signal such as the peak values of IMFs and existing sensitive spikes in the behaviors of the amplitude and energy features of the IMFs occurred owing to the stiffness changes of the structure. However, lower values of the RMS and energy feature resulting from the EMD and EEMD methods are related to the averaging operation adopted by these techniques over all realizations during their decomposition processes to produce the signal components that lead to a large variation in the number of modes. More explanations on the theoretical advantages of the CEEMDAN compared to EMD and EEMD can be found in Chapter 3.

Similar to the enhancement trend observed in the behavior of energy feature, the number, and intensity of damage spikes in the behavior of instantaneous amplitude feature extracted by CEEMDAN extremely increase with decreasing the distance of sensors relative to the damage element as shown in Figure 4.3. Although these enhancements in the *IA* results from EMD and EEMD techniques are not as much as that of CEEMDAN, the

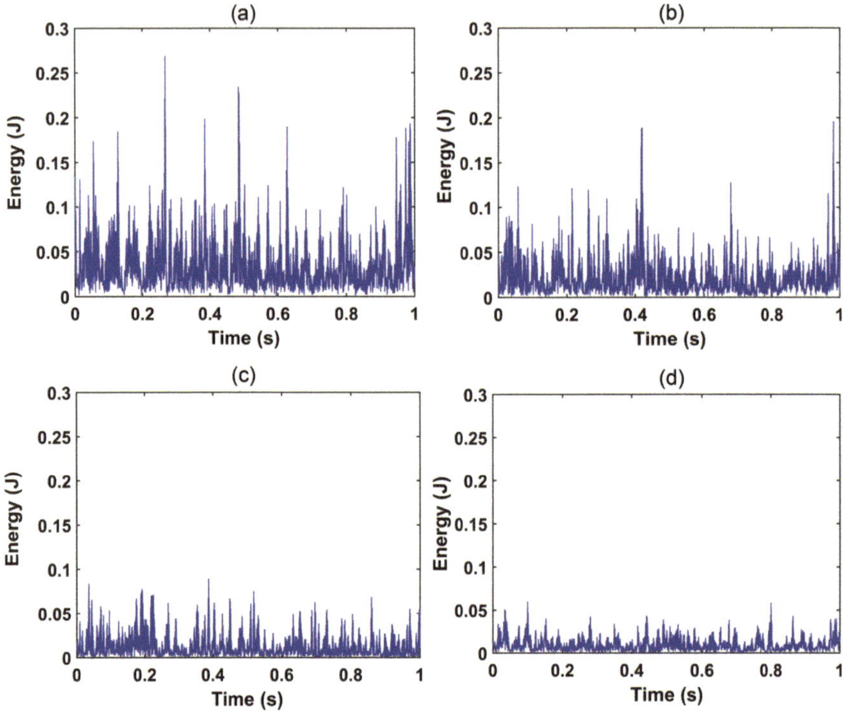

Figure 4.2 The first IMF's energy for different locations of (a) the nearest, (b) second-near, (c) far, and (d) the farthest sensors from the damaged element by the CEEMDAN technique.

increase of their magnitudes is slightly noticeable for the closer sensors. Accordingly, the magnitudes of the mean of IA and the corresponding damage index as given by Equations (2.3) and (2.4) significantly increase for the sensors located at near distances (i.e., sensors 11 and 9) relative to the damaged element compared to those of sensors located at farther distances (i.e., sensors 4 and 1) as shown in Figure 3.7a.

The damage index results defined based on the energy and instantaneous amplitude compared to their combinations with kurtosis and entropy features were presented in Figure 3.7b and c. In these figures, the sensitivities of these damage indices are illustrated not only to the damage severity (illustrated using different legends) but also to the damage location (around sensor 11) by showing the results from all 13 sensors. It is obviously seen that the damage index curves have more concentrated curvature around the damage location at sensor 11. Therefore, the results of these figures demonstrate the merit of the proposed approach based on CEEMDAN in identifying the presence, location, and severity of the damage.

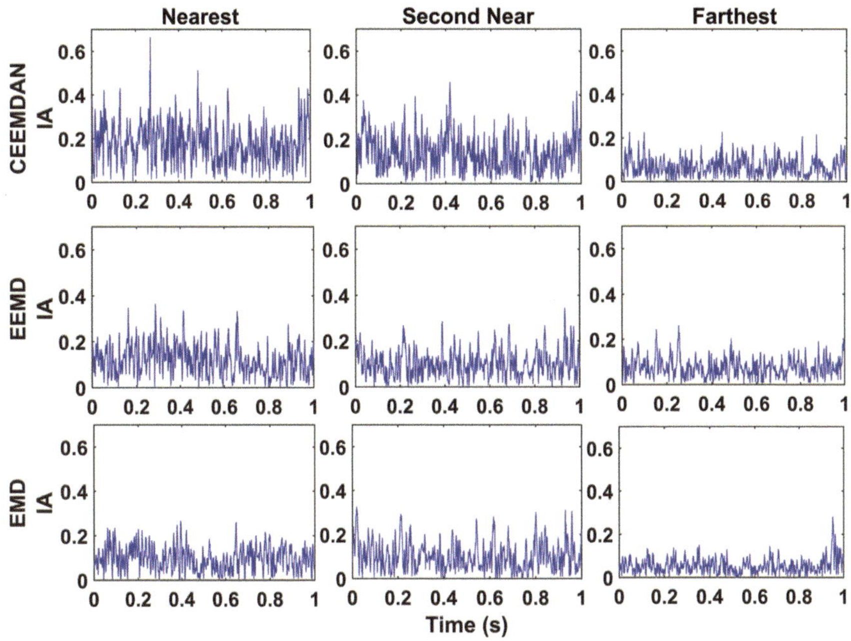

Figure 4.3 The *IA* of the first IMF results extracted by CEEMDAN, EEMD, and EMD for different sensor locations relative to the damaged element.

Besides, the unwrapped phase extracted from HHT based on CEEMDAN, EEMD, and EMD is also assessed as the indicator of the damage location. In Figure 4.4, it is observed that the deviations between damaged and healthy states increased by reducing the distance between the specified sensors and the damaged element. Therefore, increasing the angle of the unwrapped phase with the decrease of the distance of the sensors relative to the damage location results in the increase of the damage index magnitude as defined in Equation (2.6). The performance of the unwrapped phase in identifying the damage location can be also clearly seen in Figure 3.9 in which the damage index results calculated based on the unwrapped phase extracted by CEEMDAN, EEMD, and EMD techniques are plotted for all thirteen sensors. As shown in Figure 4.4, the unwrapped phase extracted by CEEMDAN is significantly more sensitive to the damage location than those of EMD and EEMD methods. Accordingly, the damage index results based on CEEMDAN show higher values and more curvature around the damage location than those of EMD and EEMD as presented in Figure 3.9). As such, the qualitative and quantitative results show considerable deviations and damage index values, respectively, from the CEEMDAN for the sensors around to the damaged element (especially for the nearest, second

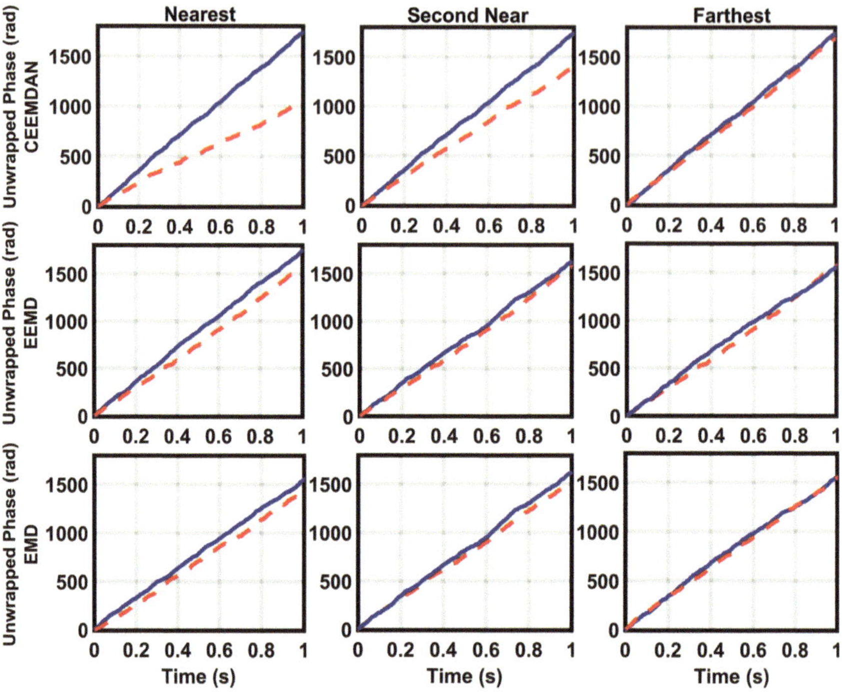

Figure 4.4 The unwrapped phase of the first IMF results extracted by CEEMDAN, EEMD, and EMD for different sensor locations relative to the damaged element.

near sensors) compared to those from far and farthest sensors which demonstrate the advantage of using the CEEMDAN in locating the damage compared to EEMD and EMD.

Detecting the location of the damage using the instantaneous frequency and corresponding the HHT spectrum, is considered as another approach in this study. This approach presents the instantaneous frequency of IMFs in a time-frequency domain. In Figures 4.5a–c, it is seen that the intensities of the IMFs increase between the frequencies of 0 and 20 Hz with the decrease of the distance of sensors from the damaged element. Upon careful observation of these figures, all four sensors show the energy density in the low-frequency spectra but it is seen that there is a decrease in the power of the frequency as the sensor location is moved farther with respect to the damage location. Generally, all four extracted features evaluated in this section were successfully able to locate the damage and demonstrate which technique has more efficient performance in classifying the damage severity and locating the damage. Consequently, the experimental results showed the capability and robustness of the CEEMDAN compared to EEMD, and EMD.

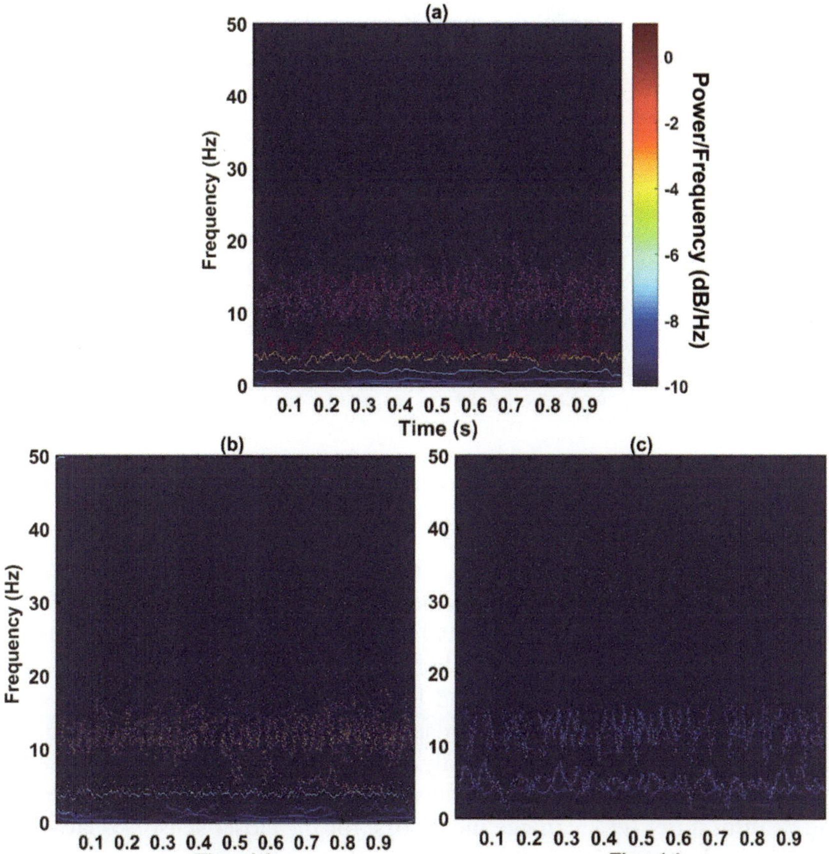

Figure 4.5 HHT spectrum for different locations of (a) the nearest, (b) second-near, and (c) the farthest sensors from the damaged element (herein, 100% damage).

4.2 DETECTION OF DAMAGE LOCATION USING CEEMAN-MUSIC TECHNIQUE

In this section, the performance of the proposed CEEMDAN-MUSIC technique in detecting the damage location is assessed by assessing the acceleration responses of the truss bridge recorded by different sensors of 11, 9, 4, and 1 that represent the sensor locations with the nearest, the second near, far, and the farthest distances, respectively, relative to the location of the damaged element. It is noteworthy that the damage scenario in this section is implemented by removing the specified diagonal element (i.e., the damage with a level of 100%). Then, the MUSIC technique is

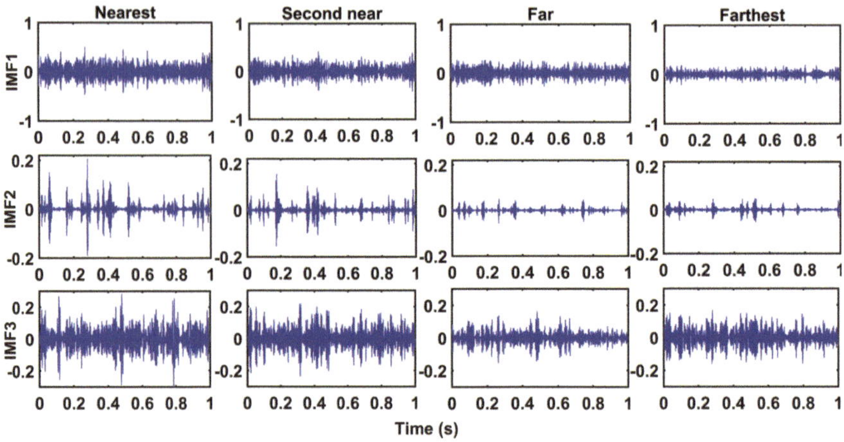

Figure 4.6 The first three IMF results extracted by CEEMDAN for the sensor locations.

applied to the acceleration response directly, and also to the IMFs extracted by CEEMDAN that have been obtained from the locations of the aforementioned four sensors on the truss bridge. Figure 4.6 illustrates the time histories of the first three IMFs obtained by CEEMDAN applied to the accelerations of the bridge recorded by the aforementioned different sensors. Figures 4.7a–d show the pseudospectral results of the bridge vibrations at the different sensor locations analyzed by the pure MUSIC technique. It is found that the distance between the pseudospectral peaks of the damaged and healthy states (i.e., damage indicators) for the closer sensors relative to the damaged element is greater than those that are far.

The pseudospectral results from the CEEMDAN-MUSIC technique in Figure 4.8a–d efficiently recognized the location of the damage. It is seen that the pseudospectrums of the first IMF were extracted from responses of four sensor locations of the truss in the healthy and the 100% damage states from when subjected to the band-limited white noise excitations. The results demonstrate that using CEEMDAN to decompose the response of the structure can improve the pure MUSIC technique in addressing the damage location. The distance between the pseudospectral peaks of the healthy and the damaged states increases with the decrease of the distance of the sensor relative to the damaged element. This means that the significant differences between the two pseudospectral are observed before and after the damage of the sensors around the damaged element, compared to those from farther sensors. Furthermore, the increases of the power peaks in the damage states are observed in proportion to the decrease in the sensor distances from the damaged element.

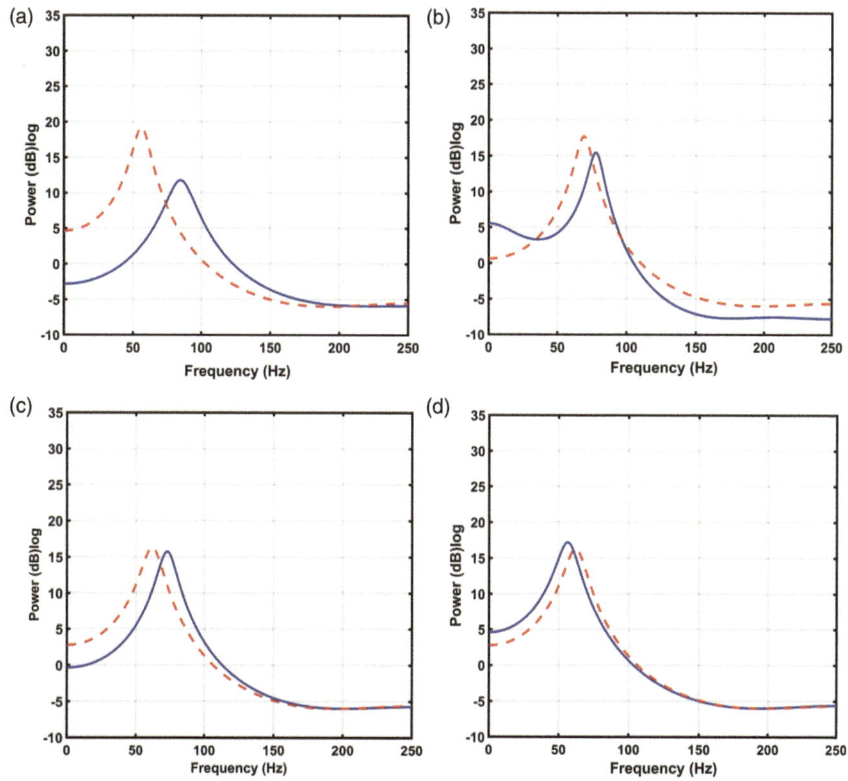

Figure 4.7 Comparison of pseudospectrum of acceleration responses by MUSIC technique for different locations with (a) the nearest, (b) the second-near, (c) far, and (d) the farthest distances relative to the damaged element.

The value of absolute differences of pseudospectral peaks in normalized frequency and power magnitude ranges are collected in Table 4.1. According to the results, it is found that the combined CEEMDAN-MUSIC method gives greater frequency and power differences in comparison to those from pure MUSIC technique. As a result of sensor-11, the values of D_f and D_p computed by CEEMDAN-MUSIC method increase by 82% and 150% compared to those of pure MUSIC method. Therefore, it can be found that the combination of CEEMDAN and MUSIC can improve the capability of the pure MUSIC technique in addressing the damage location. Besides, the differences between the peaks of pseudospectrums obtained from the healthy and damaged states of the truss are recognized as effective indicators to detect the location of damage in the structure.

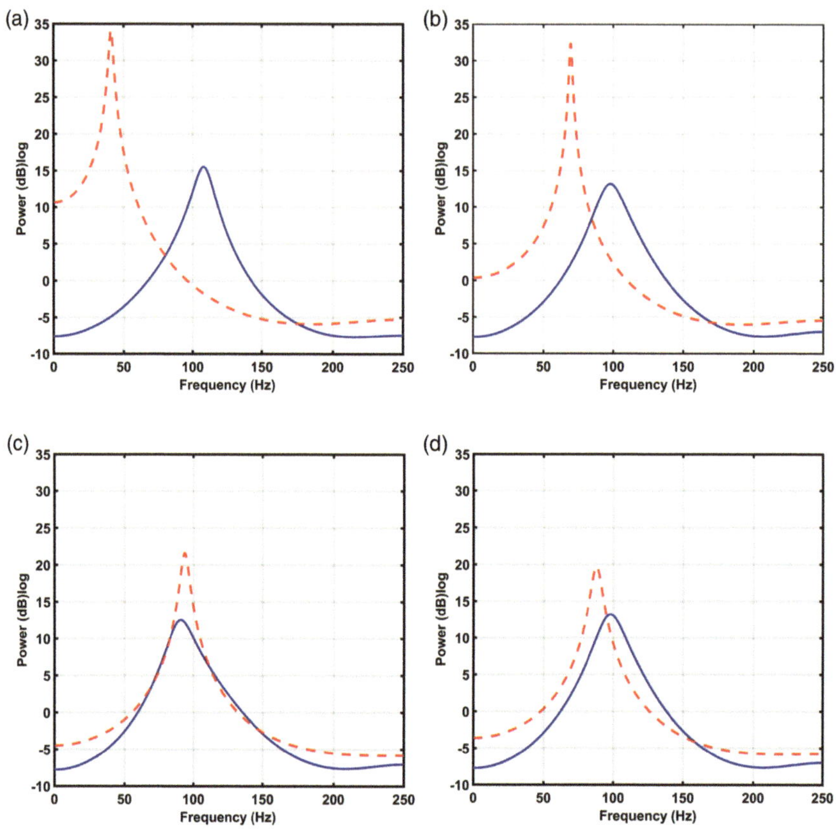

Figure 4.8 Comparison of pseudospectrum of IMF1 by CEEMDAN-MUSIC technique for different locations with (a) the nearest, (b) the second-near, (c) far, and (d) the farthest distances relative to the damaged element.

Table 4.1 The absolute difference of spectrum peaks in the normalized frequency (D_f) and power magnitude (D_p) ranges before and after damage 100% for different locations

Sensor number	*MUSIC*		*CEEMDAN-MUSIC*	
	D_f (Hz)	*D_p (dB)*	*D_f (Hz)*	*D_p (dB)*
1	0.08	0.87	0.02	9.08
4	0.13	0.57	0.11	6.52
9	0.11	2.26	0.28	19.12
11	0.34	7.42	0.62	18.62

4.3 CHAPTER SUMMARY

In the first part of this chapter, the performance of the EMD-based signal processing technique referred to as CEEMDAN was experimentally assessed in identifying the location of the damage for a steel truss bridge model. Based on the evaluations using three signal processing techniques including EMD, EEMD, and CEEMDAN, the first IMF extracted by the CEEMDAN was selected as the evaluation criterion in detecting the structural damage. In addition, four key parameters of the signal including the energy, instantaneous amplitude (*IA*), unwrapped phase, and instantaneous frequency (*IF*) extracted through applying Hilbert transform (HT) to the IMFs are considered to investigate the existence, severity, and location of the damage in the model. Furthermore, the sensitivity of CEEMDAN compared to the previous generations of the EMD-based techniques was investigated by proposing several improved damage indices from the combinations of two statistical signal features including kurtosis and entropy with the energy and instantaneous amplitude features of the analyzed signal.

The main findings and contribution of the first part of this chapter can be summarized as follows:

- The CEEMDAN approach as a novel extension of EMD accurately reproduced a proceed signal and mitigates the mode mixing problem.
- The intensity of spikes of the first IMF decomposed by CEEMDAN increased more significantly compared to those from EEMD and EMD techniques, which demonstrates the completeness of CEEMDAN in decomposing the acceleration response of the structure. Therefore, the CEEMDAN presented a more sensitive damage detection approach compared to EMD and EEMD.
- Increasing the energy and *IA* values and decreasing the unwrapped phase values of the first IMF were observed with the appearance of the damage, with increasing the level of damage, and with decreasing the distance of the desired sensor from the damage location.
- Although EMD and EEMD were able to detect the damage, a significant improvement of CEEMDAN compared to the other techniques in detecting the location of the damage using the damage indices based on energy, *IA*, and unwrapped phase parameters, was concluded.
- By assessing the *IF* of IMFs in the time-frequency-energy domain, an increase in the power of the frequency in the low ranges is observed when the damage appears. Furthermore, this technique can provide an indication of the general damage region, as well as being a sensitive indicator of damage.
- By comparing the value of the damage indices based on energy, *IA*, and unwrapped phase features, it was found that the energy feature represents a better approach to locating the severity of the damage.

- It was concluded that the proposed damage indices based on the combinations of the kurtosis and entropy features with the energy and *IA* features resulted in more sensitive indices in identifying the location of the damage compared to those based on the direct utilization of the energy and *IA* features. In addition, more sensitive indices were obtained when combining the entropy feature with the energy and *IA* compared to those in which the kurtosis feature was utilized.

The second part of this chapter experimentally assessed the performance of a hybrid CEEMDAN-MUSIC technique as a novel signal processing technique in identifying the location of damage in a laboratory-scale model of a steel truss bridge. The implementation process of this technique begins by employing the CEEMDAN algorithm to decompose the acceleration response of the truss bridge exposed to a white noise excitation. Then, the MUSIC technique was applied to the first IMF before and after damage to compute the power density pseudospectrum. In addition, the absolute differences of the spectral peaks in the frequency (D_f) and power magnitude (D_p) ranges extracted from the responses of different damage scenarios varying in terms of the damage location are considered as effective indicators to detect the location of the damage. Significant improvements were observed in the damage detection and characterization process of the truss using the proposed method by comparing the results from the indicator values of the combined CEEMDAN-MUSIC algorithm with those from pure MUSIC technique and several frequency-domain techniques such as FFT, PSD, and FDD. As such, compared to many spurious peaks captured by the conventional frequency domain techniques leading to notable difficulties in the damage detection process, the proposed method was clearly able to address the classification of the damage location, as well as the perception of damages at lower levels. The contribution of this chapter is the robustness of CEEMDAN, MUSIC, and their combination in damage detection of civil engineering structure for the first time. As a result, the method can be recommended for health monitoring of various structures according to the proposed framework in this study.

In Chapter 5, the efficiency of artificial neural network (ANN) is evaluated to automate the damage identification approach with the purpose of effectively reducing the human intervention and speeding up the process of structure diagnosis.

Chapter 5

Experimental verification of the proposed artificial neural network-aided approaches in structural damage identification

5.1 INTRODUCTION

5.1.1 Advancements in time-frequency-domain signal processing techniques

In recent decades, traditional signal processing techniques with time-frequency representation such as short-time Fourier transform (STFT) [212], wavelet transform (WT) [464], Hilbert–Huang Transform (HHT) [465] have significantly improved the damage detection process of nonlinear signals through enhancing the accuracy of feature extraction. These techniques still have some unsolved difficulties such as the inability of STFT in presenting the instantaneous frequencies with high-resolution, and the deficiency of WT in decomposing a time signal and calculating the instantaneous frequencies owing to incompatibility between the prescribed basis of WT and the analyzed signal. Although Huang et al. [466] proposed the empirical mode decomposition (EMD) technique to solve the unadaptability of WT, EMD still has some serious problems such as the mode-mixing problem and distorted components. The major disadvantages of Fourier transform (FT) method such as the low-resolution spectral representation of data, and its inefficiency in analyzing non-stationary signals were efficiently improved using several advanced techniques including Fourier decomposition method (FDM), Fourier-Bessel decomposition method (FBDM) [467], and Fourier–Bessel series expansion (FBSE) [468] techniques. Also, some improvements were applied to the aforementioned techniques by proposing FBSE-based flexible time-frequency coverage WT [469,470], and Fourier-Bessel series expansion-based empirical WT [471–476] techniques to enhance the spectral representation for multicomponent nonstationary and nonlinear signals with higher resolutions. However, these techniques are not adequately able to analyze the time-varying signals and efficiently obtain the instantaneous information of the analyzed signal simultaneously in time and frequency domains since they are frequency-based techniques.

Despite many applications of EMD in the literature, this method suffers some drawbacks such as generating undesirable IMFs at the low-frequency

DOI: 10.1201/9781003499046-5

region with the mode mixing problem leading to inaccurate SHM results. To overcome this, Torres et al. [273] proposed the complete ensemble EMD with adaptive noise (CEEMDAN) by adding a particular noise in which after each stage of the decomposition process, a unique residue is obtained.

One of the significant issues in vibration-based structural damage detection is to construct and extract the sensitive parameters from structural dynamic responses to identify the initial damages. To this end, a multifarious structural damage identification system based on the combination of signal processing and artificial intelligence techniques has been developed recently. This can guarantee the robustness, reliability, and efficiency of the detection process [477]. Since the use of the HHT method combined with neural networks is very limited in the literature, it is aimed to evaluate the performance of a proposed methodology from the combination of CEEMDAN-Hilbert transform and neural network techniques called CEEMDAN-HT-ANN in the first part of this chapter (i.e., Section 4.2), which has not been reported for damage detection in civil engineering. Due to the potential advantages of the CEEMDAN demonstrated by Torres et al. [273] compared to the previous generation of EMD-based techniques, CEEMDAN is employed to decompose the acceleration response of a steel-truss bridge model, which was experimentally established in laboratory settings and subjected to a band-limited white noise excitation. The vast majority of previous research studies used modal parameters for structural health monitoring. Besides, evaluating the signal features in both time and frequency domains all at once is a gap of knowledge in SHM and damage identification in civil engineering applications especially for complex structures such as bridges. The change in structure stiffness that occurred due to structural damage would substantially cause potential changes and unexpected anomalies in these features. Therefore, it aims to qualitatively and quantitatively assess the sensitivity of these features to the presence, location, and severity of the damage in the truss as the evaluation criteria for damage detection in this study. The main advantage of this study is the use of four signal features including energy, IA, unwrapped phase, and IF that extracted through applying the Hilbert transform on decomposed IMFs by CEEMDAN. These features have not been evaluated all at once in damage detection procedure in civil engineering applications. Indeed, from an extensive review on the evaluation of these features, it was concluded that the change of the structure stiffness (i.e., the occurrence of damage) extremely affects the output results based on the aforementioned features. Therefore, these signal features are utilized in damage evaluation and identification through the proposed methodology. In addition, adopting ANN as a supervised machine learning with the purpose of automatic signal analysis and damage identification can effectively speed up the process of structure diagnosis and reducing the human factor.

The uniqueness and special advantages of using the HHT based on CEEMDAN compared to conventional techniques such as FFT or wavelet is the feasibility of analyzing the nonlinear and nonstationary data and extracting the key features of the decomposed signal including IA, energy, unwrapped phase, and IF. These characteristics of IMFs can reflect how the energy and phase of the signal vary with time. Moreover, the IF feature indicates the data in a time-frequency-power domain through the spectrogram plots. Besides, since the IMF is almost mono-component, all the instantaneous parameters from a nonlinear and non-stationary signal can be efficiently captured by HT. The information obtained by these features can provide a more accurate and real-life representation of the signal which can overcome some shortcomings such as artifacts associated with the nonlocal and adaptive limitations generated by FFT and wavelet methodologies.

Besides, to overcome the limitations associated with the aforementioned signal processing techniques, Gilles [476] proposed the algorithm of the empirical wavelet transform (EWT) technique in which nonlinear signals are efficiently represented by time-frequency functions and instantaneous information of the analyzed signal can be adequately obtained in both time and frequency domains. The main advantages of utilizing EWT are to improve the frequency resolution limitations of data associated with the previous techniques, and efficiently separating the signal components in which the ratio of the signal low frequency to its high frequency is greater than 0.75. The EWT is a popular adaptive decomposition algorithm that has an excellent capability in analyzing non-linear and non-stationary signals. This technique has been widely used in the literature to accurately extract the components of the Fourier spectrum of signals. However, EWT has a limitation in decomposing a noisy signal which is composed of different segments of the Fourier spectrum without any distinct separation between the segments (i.e., existing significant overlap between the segments). Therefore, selecting an optimal approach to segmenting the Fourier spectrum components of a signal with correct boundaries plays a key role in obtaining the correct decomposition results by EWT.

Compared to the extensive use of the EWT technique for fault diagnosis of mechanical systems, it has been rarely used in damage detection of civil structures. Besides, EWT has been mostly employed to identify the modal parameters and frequency-domain features of signals. The first step in the feature extraction process is the transformation of vibration signal from time to frequency domain, which is commonly implemented using the FFT technique. However, FFT is not effectively able to process nonlinear signals. Hence, the analysis of time-domain features of a nonstationary signal would be more reasonable and successful. Owing to the advantages of the EWT in obtaining more improved data with high-resolution in both time and frequency domains and its high capability in decomposing nonlinear and nonstationary signals compared to conventional techniques mentioned in Section 5.1.1, EWT is adopted in this study to decompose the signal responses of a laboratory-scale model of a steel truss bridge under different excitations.

Due to existing drawbacks of traditional modal-based techniques such as FFT in obtaining time-dependent instantaneous information of signals, several time-domain statistical features including RMS, shape factor, kurtosis, and entropy are employed in this study, which are efficiently able in fault diagnosis. In particular, the kurtosis and entropy features are more effective in the analysis of nonlinear signals since they are operated based on the analysis of probability density function (PDF) of signals which can present the information of discrete-time signal segments [3,402]. Hence, time-domain statistical features (especially kurtosis and entropy features) are highly sensitive to sudden stiffness changes representing the occurrence of structural damages. Therefore, this study aims to extract the time-domain features of the signal components decomposed by EWT.

5.1.2 An overview of artificial intelligence

Artificial neural networks have emerged as a focal point of extensive study, aiming to emulate human-like efficiency in computational systems. Comprising both linear and nonlinear computational elements working in concert, these networks represent cutting-edge advancements in machine learning, knowledge representation, and the practical application of acquired knowledge to predict outcomes in complex systems.

The foundational idea behind neural networks draws inspiration, to some extent, from the intricate processes of data and information processing observed in biological neural systems. The core concept involves the development of innovative structures for information processing, wherein numerous highly interconnected processing elements, referred to as neurons, collaborate to address problems and facilitate the transfer of information through synapses, akin to electromagnetic communications. Remarkably resilient, if a cell within these networks becomes damaged, other cells can compensate for its absence, contributing to the network's adaptive reconstruction. One of the remarkable capabilities of these networks is their capacity for learning. Drawing parallels from the biological realm, such as the burn response in touch nerve cells, artificial neural networks showcase their ability to adapt and learn from experiences. When a cell is subjected to a burn, it learns not to approach hot objects, and the system, guided by algorithms, corrects itself. The learning process is comparative; as new inputs are introduced, the weights of synapses undergo adjustments, enabling the system to generate increasingly accurate responses. In essence, artificial neural networks stand at the forefront of innovation, mirroring biological principles to create intelligent systems capable of dynamic learning and problem-solving.

The definition of a neural network lacks a universal consensus among researchers; however, a prevailing viewpoint posits it as a network composed of simple processing elements, commonly known as neurons. These neurons exhibit a complex behavior determined by the intricate relationships between processing elements and their associated parameters. The inspiration for this computational technique is drawn from the examination of the central nervous system and biological neurons—specifically, the axons, multiple branches of nerve cells, and junctions of nerves—integral components of information processing in the nervous system.

In a neural network model, simple nodes or processing elements are interconnected, forming a network of nodes, and this interconnected structure is why it is aptly named "neural network". While a neural network may not possess inherent adaptability, it becomes practically applicable through specialized algorithms designed to alter the communication weights within the network, thereby generating desired signals. The translation of neural network concepts into a functional data structure resembling a neuron is achievable through computer programming. Training the network involves creating a network of interconnected artificial neurons, developing a training algorithm, and applying this algorithm to the network. Pioneering work in neural networks dates back to the late 1940s, with McCulloch and Pitts exploring the interconnection capabilities of a neuron's model. They proposed a computational model based on a basic neuron-like element. In 1949, Donald O. Hebb introduced a learning rule to adapt connections between artificial neurons. Subsequently, in 1958, Rosenblatt contributed to the field with the perceptron algorithm and the statistical separability theory. This historical progression underscores the evolution of neural network research, from foundational concepts of neuron-like elements to the development of learning rules and algorithms, laying the groundwork for the sophisticated neural networks employed in contemporary machine learning and artificial intelligence applications.

5.1.2.1 Applications of neural networks

Neural networks find diverse applications across various fields due to their ability to process complex information and adapt to different tasks. Some prominent applications include:

1. Pattern Recognition: Facial recognition, fingerprint recognition, voice and speech recognition, and handwriting recognition are common applications. For instance, banks use neural network chips to compare signatures for cash withdrawals.
2. Medicine: Neural networks are utilized in electrocardiogram signal analysis, disease diagnosis, and prescription recommendations.
3. Commercial Applications: Neural networks assist in complex decision-making processes requiring extensive data, such as predicting stock fluctuations in the stock exchange market based on historical data.

4. Artificial Intelligence (AI): Many AI experts consider artificial neural networks crucial for designing intelligent machines.
5. Visual Data Compression: Neural networks contribute to data reduction through visual data compression.
6. Noise Elimination on Telecommunication Lines: Neural networks are employed to eliminate noise on telecommunication lines.
7. Military Systems: Applications include submarine mine detection and eliminating abnormal sounds in radar tracking systems.
8. Constructing and Operating Building Structures: Neural networks expedite the process of determining optimal building structures through high-speed data processing and analysis.
9. Marketing: Online advertising utilizes neural networks to enhance and optimize sales strategies.
10. Monitoring: Danger prediction in spacecraft and analysis of sounds produced by diesel engines in rails are examples of neural network applications in monitoring systems.
11. Other Applications: Risk analysis systems, pilot-less aircraft control, welding quality analysis, computer quality analysis, emergency room (ER) testing, oil and gas exploration, truck braking detection systems, loan risk estimation, spectral recognition, medication detection, industrial control processes, error management, sound recognition, hepatitis diagnosis, remote data retrieval, 3D object recognition, and more.

Neural network applications can be categorized into correspondence, clustering, categorization, identification, pattern reconstruction, generalization, and optimization. They are increasingly used in tasks such as pattern recognition (e.g., handwriting, speech, and visual recognition) and controlling or modeling systems with unknown or complex internal structures, as seen in engine input control. The versatility of neural networks makes them a valuable tool in addressing a wide range of challenges across various domains.

5.1.2.2 Neural network training algorithms

The learning process in neural networks can be categorized into different types, each suited for specific tasks:

1. Supervised Learning:

   This approach focuses on a specific subject, providing the network with a variety of examples. The network analyzes both the input data and examples, eventually gaining the ability to recognize new types of examples it has not encountered before. This is a commonly used method.
2. Unsupervised Learning:

   Unsupervised learning involves a higher level of learning and is less commonly utilized in comparison to supervised learning. In this approach, the network is not provided with labeled examples, and it must find patterns and relationships within the data on its own.

3. Reinforcement Learning:
   This type of learning involves an agent interacting with an environment and learning to make decisions to achieve a goal. Reinforcement learning often utilizes models such as Hidden-Mode Markov Decision Processes (HM-MDPs), which include states, actions, transitions, and the added value of each action.
4. Hidden-Mode Markov Decision Processes (HM-MDPs):
   HM-MDPs are a specific model within reinforcement learning, comprising states, actions, transitions, and instantaneous added values for actions. This model is particularly relevant in scenarios where an agent needs to make decisions over time.

For neural networks, certain problems are more suitable for effective learning. Some considerations include:

- Error in Training Data: Neural networks can handle training data with errors, such as noise from sensor data collected from different devices like cameras and microphones.
- Objective Function with Continuous Values: Neural networks are well-suited for problems where the objective function has continuous values.
- Sufficient Learning Time: Learning in neural networks may require more time compared to other techniques, such as decision trees. It is important to have sufficient time for the network to learn effectively.
- Stable Objective Function: Neural networks are advantageous when the objective function does not need frequent changes. Changing the weights learned by the network can be a complex task, making neural networks suitable for scenarios with a stable objective function.

Understanding these considerations helps in selecting the appropriate learning approach and methodology based on the characteristics of the data and the requirements of the task at hand.

There are several theories and technologies that contribute to the field of artificial intelligence. Here are some of the main ones:

1. Neural networks (NNs): these structures have a good ability to find a model for data and analyze the data. NNs have wide applications in many branches of science and technology. Many structures have been proposed for various applications in computer vision, engineering, medical science, social problems, and economic problems. The most famous NNs are listed as follows:
   - Multi-layer perceptron (MLP)
   - Convolutional Neural Networks (CNN)
   - Long Short-Term Memory (LSTM) Networks

- Restricted Boltzmann Machine
- Group Method of Data Handling (GMDH) neural networks
- Cascade-Correlation Adaptive Method (CMAC) neural networks
- Gated Recurrent Unit (GRU)
- Bidirectional Recurrent Neural Networks (BRNN)
- Echo State Networks (ESN)
- Kohonen Self-Organizing Maps (SOM)
- Adaptive Resonance Theory (ART) Networks
- Growing Neural Gas (GNG)
- Neural Gas (NG)
- Counterpropagation Networks (CPN)
- Recursive Neural Networks (RNN)
- Recurrent Autoencoders (RAE)
- Generative Adversarial Networks (GAN)
- Spiking Neural Networks (SNN)
- Liquid State Machines (LSM)
- Recursive Neural Networks (RNN)
- Discrete Hopfield Network
- Continuous Hopfield Network
- Binary Hopfield Network
- Stochastic Hopfield Network
- Asynchronous Hopfield Network
- Bidirectional Associative Memory (BAM) Hopfield Network
- Content-Addressable Memory (CAM) Hopfield Network
- Restricted Boltzmann Machine (RBM) Hopfield Network
- Recurrent Convolutional Neural Network (RCNN)
- Deep Convolutional Neural Network (DCNN)
- Convolutional Restricted Boltzmann Machine (CRBM)
- Spatial Transformer Network (STN)
- Fully Convolutional Network (FCN)
- Residual Network (ResNet)
- Inception Network (InceptionNet)
- U-Net
- Mask R-CNN

2. Fuzzy logic systems (FLSs): FLSs present a powerful tool to represent linguistic variables, and model uncertain data. FLSs also have gotten great attention in many engineering applications.
   - Interval Type-2 Fuzzy Systems (IT2FS)
   - General Type-2 Fuzzy Systems (GT2FS)
   - Simplified Type-2 Fuzzy Systems (ST2FS)
   - Hybrid Type-2 Fuzzy Systems (HT2FS)
   - Adaptive Type-2 Fuzzy Systems (AT2FS)
   - Evolving Type-2 Fuzzy Systems (ET2FS)
   - Interval Type-2 Fuzzy Neural Networks (IT2FNN)

- Type-2 Fuzzy Logic Controllers (T2FLC) Hybrid fuzzy system
- Fuzzy decision-making system
- Fuzzy clustering system
- Type-1 fuzzy control system
- Fuzzy expert system
- Fuzzy neural networks
- Fuzzy rule-based system
- Fuzzy inference system with multiple inputs and outputs (MIMO-FIS)
- Type-3 fuzzy logic system
- Self-organizing Fuzzy neural networks (SOFNN)
- Self-organizing Fuzzy logic controllers (SOFLC)
- Self-tuning Fuzzy controllers (STFC)
- Adaptive Fuzzy systems
- Self-adaptive Fuzzy inference systems (SAFIS)
- Hierarchical Fuzzy systems
- Evolving Fuzzy systems
- Incremental Fuzzy systems
- Dynamic Fuzzy systems
- Online Fuzzy systems

3. Learning algorithms (LAs): These techniques allow FLSs and NNs to learn from data. LAs help intelligent systems learn the pattern of data and used for modeling, controlling, analyzing, and forecasting. LAs include:
   - Evolutionary-based learning schemes such as

   1. Genetic algorithms (GA) 2. Particle swarm optimization (PSO) 3. Differential evolution (DE) 4. Ant colony optimization (ACO) 5. Artificial bee colony (ABC) 6. Cultural algorithms (CA) 7. Evolutionary programming (EP) 8. Evolution strategies (ES) 9. Genetic programming (GP) 10. Harmony search (HS)
   - Basic optimization techniques such as

   1. Kalman filter 2. Gradient descent 3. Newton's method 4. Broyden-Fletcher-Goldfarb-Shanno (BFGS) algorithm 5. Conjugate gradient method 6. Nelder-Mead method 7. Simulated annealing 8. Hill climbing 9. Stochastic gradient descent 10. Levenberg-Marquardt algorithm
   - Unsupervised learning such as

   1. Clustering algorithms (e.g. k-means, hierarchical clustering) 2. Principal component analysis (PCA) 3. Independent component analysis (ICA) 4. Autoencoders 5. Generative adversarial networks (GANs) 6. t-distributed stochastic neighbor embedding (t-SNE) 7. Association rule learning (e.g. Apriori algorithm) 8. Neural network architectures for unsupervised learning (e.g. Boltzmann machines, deep belief networks)

- Semi-supervised learning such as
  1. Self-training 2. Co-training 3. Multi-view learning 4. Graph-based methods 5. Semi-supervised support vector machines (SVM) 6. Label propagation 7. Expectation-maximization (EM) algorithm 8. Generative models (e.g. Gaussian mixture models) 9. Active learning with unlabeled data 10. Transfer learning with labeled and unlabeled data
- Deep learning
- Transfer learning
- Online learning
- Ensemble learning
- Bayesian learning
- Instance-based learning

4. Natural Language Processing (NLP): This is a special case of application of NNs, FLSs, and LAs that focuses on understanding and interpreting the human language. NLP, I used for text and sentiment analysis, and speech recognition.
5. Robotics: Robotics can be considered as a special case of application of NNs, FLSs, and ALs in mechatronic field. Because of the wide research in this area, we can consider it as a main field of research in AI.

Since structural health monitoring of complex civil structures such as bridges in which damage detection and localization are confusing due to existing big data for analysis, a vital need for automated structural damage assessment systems to undertake the soundness, validity, and efficiency of the diagnostic process, is the topic of importance. The main purpose of utilizing Artificial Neural Network (ANN) in this study is to automate the damage detection procedure based on EWT in a reliable, quick, and simple manner for a complex truss bridge structure. This automation provides an intelligent decision-making procedure to accurately detect the presence and location of the damage in complex civil structures such as truss bridges in which data is quite big.

Despite different combinations of EWT with different machine learning algorithms, there exists no hybrid damage detection approach reported in the literature from the combination of EWT and ANNs to detect the damage in civil engineering structures using time-domain statistical features. The main advantage of ANNs compared to other machine learning algorithms is that they allow the nonlinear and non-parametric modeling of complex systems with large datasets simply and adaptively without requiring explicit mathematical representations or without requiring exhaustive experiments. However, most other statistical methods are parametric models that need a high background in statistics. Hence, the use of ANN and its hybridization with signal processing techniques are more popular for structural health monitoring (SHM) applications compared to other intelligent algorithms.

Therefore, Section 5.3 of this chapter proposes an automated damage detection approach from the combination of EWT and ANN techniques (named EWT-ANN technique) based on several time-domain statistical features, which is a research gap of previous works. The validity of a novel damage detection approach based on a combination of the EWT algorithm and an ANN named EWT-ANN is investigated in this study to identify the damage existence and find its location in a laboratory-scale model of a steel truss bridge exposed to band-limited white noise and impact excitations. The ANN is used in this study to automate the damage detection procedure based on the EWT and provide relationships between the signal modes as the inputs and signal features as the target outputs of the baseline state (healthy). After training the EWT-ANN damage detection model based on the signal modes and features captured from the healthy state of the bridge, the signal modes from the damage scenarios are tested compared to the trained model to automatically extract the signal features of the damage scenarios and calculate the damage indices inside the EWT-ANN model. The proposed damage detection approach quantitatively performs using several time-domain damage indices based on the signal energy feature and a series of statistical time-domain features including RMS, shape factor, kurtosis, and entropy that have been rarely used in the damage detection process of complex civil structures such as bridges [478–480].

In this chapter, first, the applications of ANN in damage detection of civil engineering structures and also its hybrid use with signal processing techniques are introduced in Section 5.2. Then, Section 5.3 of this chapter evaluates the performance of a hybrid damage detection approach combined with CEEMDAN and ANN techniques. In addition, the performance of another automated damage identification approach defined based on the combination of EWT and ANN techniques is assessed in Section 5.4.

5.2 APPLICATION OF ANN

Artificial Neural Network (ANN) has been widely used in SHM to automate the damage identification approach with the purpose of effectively reducing the human intervention and speed-up the process of structure diagnosis [481]. Most of the previous research works adopted the structural response parameters to train the ANN for the identification of the damages in civil engineering structures [176,482–484]. An ANN-based structural damage detection approach was presented by Masri et al. [485] to monitor the linear and nonlinear systems through structural parameters even in the presence of noisy environments. Thereafter, Masri et al. [486] proposed a nonparametric damage detection approach based on nonlinear identification methods for the structural unknown system through the ANN model. Dackermann et al. [487] utilized cepstrum analysis and artificial neural networks to determine the structural parameter relying on response-only measurements and damage detection procedures, respectively. Qian and Mita [484] proposed an acceleration-based damage evaluation methodology for a building structure subjected to various excitation and damage levels using ANN to detect the location and severity of the damage. The performance of ANN was assessed by the Relapsing-Remitting Multiple Sclerosis (RRMS) damage indices and its effectiveness was concluded. Arangio and Bontempi [488] utilized Bayesian ANN based on mode shape and frequency features for long-term monitoring and damage identification of a cable-stayed bridge as a real benchmark case study. The capability of the proposed method in detecting the damage was found compared to the traditional vibration-based techniques. Zang et al. [489] assessed the advantages of combining independent component analysis (ICA) and ANN in damage detection procedure of two example models of truss and three-story bookshelf structure. Experimental results demonstrated the effectiveness of the proposed methodology.

One of the significant issues in vibration-based structural damage detection is to construct and extract the sensitive parameters from structural dynamic responses to identify the initial damages. To this end, a multifarious structural damage identification system based on the combination of signal processing and artificial intelligence techniques has been developed recently. This can guarantee the robustness, reliability, and efficiency of the detection process [477]. Garcia-Perez et al. [490] combined wavelet packet transform (WPT) and EMD techniques and utilized ANN to locate and detect the combined damage of a five-bay truss-type structure. Xun and Yan [491] evaluated the capability of the combination of radial basis function neural network (RBFNN) and HHT in vibration signal analysis. RBFNN is used as a pre-processor to extend the length of the signal for removing the end swings problem and as a post-processor to select the optimal IMFs.

Conventional classification techniques such as Autoregressive Integration Moving Average (ARIMA) and Exponential smoothing (ES) are fitted to time series data either to better understand the data or to predict future points in the series (forecasting). Basically, the forecasting process depends on the length of the series and properties. If time-series data set are short and have a trend, then, ARIMA or ES can be used as a classical method. These classical techniques are suitable to analyze the short data set and take care of trends, seasonality, and cycles. These techniques are commonly used for the forecast in finance and economics. For long time-series with seasonality, Artificial Neural Network is used as a powerful machine learning method. ANN has been a better choice for researchers in structural and tools condition monitoring because of its advantages such as superior learning, noise suppression, and parallel computation. One of the limitations associated with the ARIMA model is that it is not able to capture the nonlinear pattern of the time series variable reflected in the nonlinear and non-stationary responses of structures in the real world [492]. Besides, the ES method is only a class of linear model which can only capture the linear feature of time series. Since the time series of the responses of real-world structures are often full of nonlinearity and irregularity, therefore, the ES model is unable to find accurate nonlinear patterns in the time series data [224]. In SHM and damage diagnosis of complex structures such as bridges in which data sets are long and big, ANNs are flexible and suitable machine learning to speed-up the damage diagnosis process. In addition, they enhance the ability of damage detection procedure to learn by examples which make them very flexible and powerful. Through a comprehensive study on machine learning techniques existing in the literature, ANN is more compatible with the extracted data set of the truss.

A multilayer feed-forward neural network is an interconnection of perceptron in which data and calculations flow in a single direction, from the input data to the outputs. The number of layers in a neural network is the number of layers of the perceptron. Fundamentally, feed-forward models of ANN are the first and simplest type of ANN and commonly used for damage identification of civil structures which take fixed sizes for input and output data. Although a recurrent neural network (RNN) [493] is a type of ANN, it is applicable in case the input/output size is variant. In this study, since the length of the input/output data set is fixed on 2000 data points, therefore, ANN is adopted as a supervised machine learning with the purpose of automatic damage identification, optimization, control, and prediction. The use of ANN can effectively speed up the process of structural damage diagnosis and reduce the human factor. Employment of such neural network techniques allows predicting the structural health during the service without the need to interrupt or terminate the usage of the structure.

The basic idea of using ANN as a supervised machine learning tool in this study is to establish an artificial intelligence component in identifying the structural damage through its pattern recognition. The training procedure of the learning machine is on the basis of the iterative calculation of parameters identified in the network to reach the converged outputs by minimizing the compute error between the target and obtained network outputs. The training test in the network is carried out based on the prediction process for the new input data set which is called "network generalization". To this end, a feed-forward multilayered perceptron (FFMLP) architecture is used to predict the presence of damage in the truss bridge. The multi-layer perceptron (MLP) architecture includes an input layer, one or more hidden layers, and an output layer each of which contains a set of various nodes that learned using the Levenberg-Marquardt algorithm. In this study, feed-forward networks include 20 hidden layers in which the input data are trained through the backpropagation algorithm. To do this, the Neural Networks Toolbox of MATLAB is used to implement the ANN simulation. In order to activate the neurons in the hidden and input layers, a log-sigmoid transfer function written as $f(N)=1/[1+\exp(-N)]$ is used in which N is the number of neurons in layer. In addition, the *trainlm* function is used according to the resilient backpropagation algorithm for training of the networks. The accuracy of the neural network utilized in this study is examined in the following sections using the Mean Square Error (MSE) method.

Before using the FFMLP ANN to diagnose the health condition of the bridge, first, the acceleration response of the truss before and after damage scenarios are decomposed through CEEMDAN to generate IMFs. Then, the Hilbert transform applied to each IMF to extract key parameters including instantaneous amplitude (IA), energy, unwrapped phase, and IF. The obtained IMFs and the four aforementioned features are respectively used as the input and target layers to train the ANN in a healthy state as a reference condition of the truss. That is, the training process of the ANN is based on the healthy state of the truss bridge. The sizes of the input, hidden, and output nodes in the proposed ANN are 10, 20, and 4, respectively. In implementing this methodology, IMFs 1–10 decomposed by CEEMDAN for each of 13 sensors, are selected as the inputs to the network and four extracted parameters including energy, IA, unwrapped phase, and IF are selected as the outputs of the network. Finally, the extracted IMFs from the acceleration responses of damaged truss through various damage scenarios are used to test the trained ANN.

5.3 HYBRID APPLICATION OF CEEMDAN AND ANN TECHNIQUES

In this section, an efficient damage detection approach is introduced through the combination of the vibration signal processing method and neural network. To perform the CEEMDAN-HT-ANN model in damage detection, at first, the vibrations of the steel truss bridge model subjected to a band-limited white noise in the healthy and damaged states are acquired from the identified accelerometer sensors as shown in Figure 5.1 [224]. Afterward, the vibration responses are decomposed using the CEEMDAN to generate the set of IMFs. Then, the HT is applied to the IMFs to extract their features including IA, energy, unwrapped phase, and instantaneous frequency (IF). After employing the HHT, a multi-layer perceptron (MLP) neural network is defined to train the relationship between IMFs as the input layers and their four aforementioned features for the healthy state of the bridge as the output layers. The ANN is separately trained based on each feature before applying the damage scenarios on the bridge. The outputs of the CEEMDAN-HT-ANN model based on four aforementioned features extracted from the healthy state of the bridge are compared with those from the scenarios varying in terms of the damage level, and the sensor location relative to the damaged element to classify the severity and detect the location of damage, respectively. Finally, the outputs of the CEEMDAN-HT-ANN model based on IA, energy, unwrapped phase, and

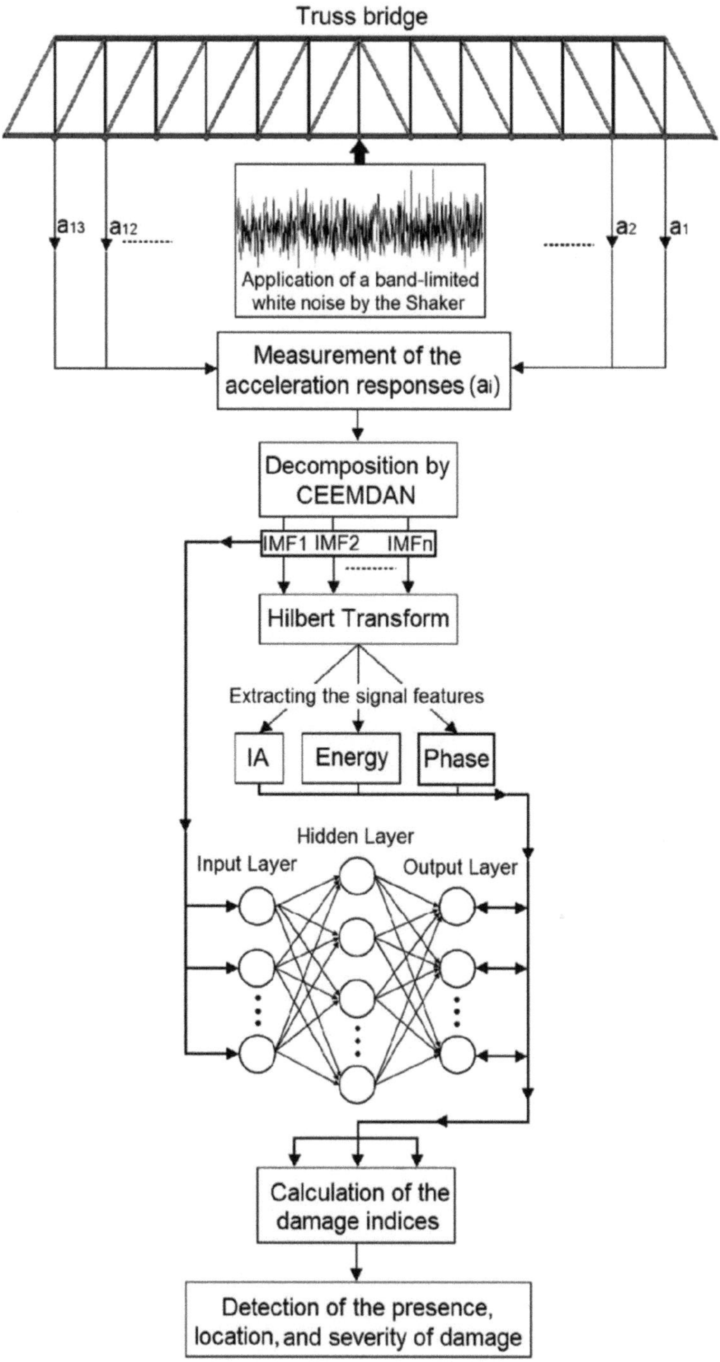

Figure 5.1 The framework flowchart of the proposed CEEMDAN-HT-ANN model to detect, locate, and classify the severity of the damage.

IF are evaluated before and after damage, scenarios using different damage indices based on the features captured from the healthy and damaged states of the truss bridge. The defined damage indices indicate the presence, location, and severity of the damage. The framework flowchart of the proposed methodology is shown in Figure 5.1 and further explanations of this flowchart can be found in Section 5.4.

5.3.1 Damage indices based on CEEMDAN-HT-ANN model

After training and testing the CEEMDAN-HT-ANN model as described in detail in Section 2.3, the differences between the outputs of the model for healthy and damaged states of the truss indicate the presence of the damage based on IA, energy, unwrapped phase, and IF parameters. In this study, three different damage indices (DIs) based on energy, IA, and unwrapped phase are defined to detect, locate, and classify the severity of damage in a steel-truss bridge to compare the differences between the outputs of the CEEMDAN-HT-ANN model for the healthy and damaged states. It should be noted that the high scalar values of DIs represent the existence of the damage. Similarly, for locating the damage on the truss, the higher values of DIs for the sensors that are near to the location of the damaged element indicate the location of the damage. In addition, it is expected to increase the damage index values by increasing the damage severity.

The energy (E) of signals is defined as the sum of the square of the IMFs given as follows:

$$E = \int_{0}^{t_0} (\text{IMF})^2 dt \tag{5.1}$$

where the parameter E is selected as one of the parameters in the output layer.

The IMFs and the energy are used as the input and target layers for the healthy state of the truss in the training process of the ANN, respectively. Thereafter, the IMFs extracted from the acceleration responses of the damage scenarios are used to test the ANN model. Then, the energy-based damage index compares the differences between the outputs of the CEEMDAN-HT-ANN model for the healthy and damaged states defined as follows:

$$DI_{(E)} = \left| \frac{E_{\text{Healthy}} - E_{\text{Damaged}}}{E_{\text{Healthy}}} \right| \times 100 \tag{5.2}$$

E_{Healthy} and E_{Damaged} represent the output of the CEEMDAN-HT-ANN model based on the acceleration response of the healthy and damaged states, respectively.

A similar procedure as defined for the energy feature is employed on the IA parameter extracted through applying of Hilbert transform (HT) to each IMF and calculating the average of the IMFs given as follows:

$$\overline{IA}_{\mathrm{IMF}} = \frac{1}{n}\sum_{i=1}^{n} \overline{IA}_i \tag{5.3}$$

where $\overline{IA}_i = \left[\overline{IA}_1, \overline{IA}_2, \ldots, \overline{IA}_n\right]$ represents the IA of each IMF extracted by HT, and n is the number of time samples. The IMFs and IA_{IMF} are used as the input and target layers of a healthy state of the truss in the training process of the ANN, respectively. Then, the IMFs of the damage scenarios are used to test the ANN. Accordingly, the damage index based on IA can be defined as follows:

$$DI_{(IA)} = \left|\frac{\overline{IA}_{\mathrm{Healthy}} - \overline{IA}_{\mathrm{Damaged}}}{\overline{IA}_{\mathrm{Healthy}}}\right| \times 100 \tag{5.4}$$

IA_{Healthy} and IA_{Damaged} represent the output of the CEEMDAN-HT-ANN model based on the acceleration response of the healthy and damaged states, respectively.

As discussed in Chapter 2, $\theta(t)$ is the phase of the IMFs are calculated by Equation (5.5). The same procedure is carried out for the unwrapped phase parameter (P) extracted by applying the Hilbert transform (HT) to each IMF and calculating the average of the IMFs given as follows:

$$\overline{P}_{\mathrm{IMF}} = \frac{1}{n}\sum_{i=1}^{n} P_i \tag{5.5}$$

where $P_i = \left[P_1, P_2, \ldots, P_n\right]$ denotes the unwrapped phase of each IMF extracted by HT.

The IMFs and P_{IMF} are used as input and target layers of the healthy state of the truss in the training process of the ANN, respectively, then the IMFs of the damage scenarios are used to test the ANN. Accordingly, the damage index based on P can be defined as follows:

$$DI_{(P)} = \left|\frac{P_{\mathrm{Healthy}} - P_{\mathrm{Damaged}}}{P_{\mathrm{Healthy}}}\right| \times 100 \tag{5.6}$$

P_{Healthy} and P_{Damaged} represent the output of the CEEMDAN-HT-ANN model based on the acceleration response of the healthy and damaged states, respectively.

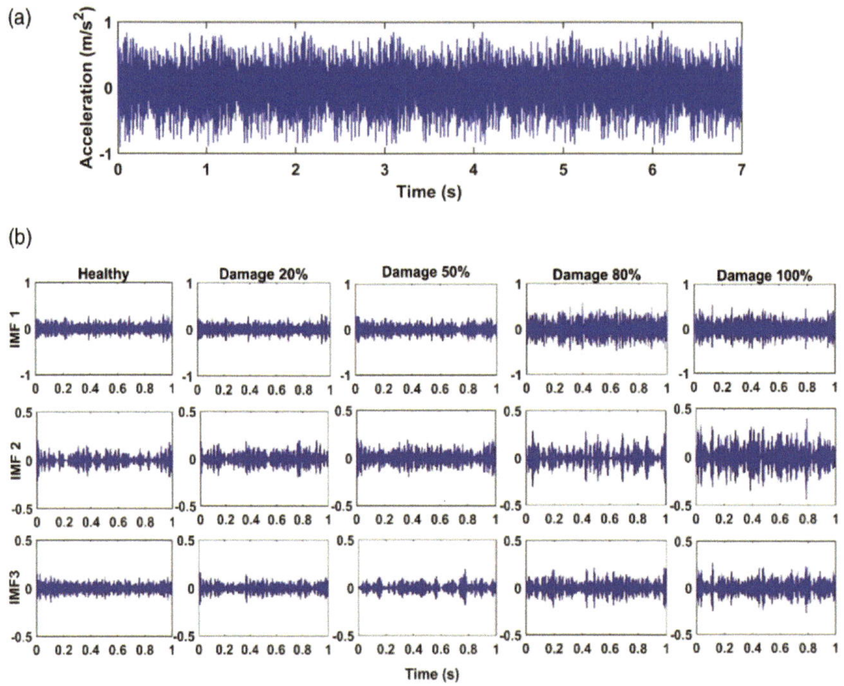

Figure 5.2 (a) Acceleration response of sensor-10 for a healthy state of the bridge, (b) the first three IMFs extracted by the CEEMDAN for sensor-10 for healthy state and different damage levels.

5.3.2 Detection of the presence and severity of damage

In this section, the presence and severity of the damage are studied by investigating the results from different damaged states (damage scenarios) of the truss bridge as described in Chapter 2. To this end, sensor-10 is selected as an example checkpoint located nearby the damaged element. Figure 5.2a and b show a segment of acceleration response of sensor-10 with a duration of 7.0 s before damage (i.e., healthy state), and the first three IMFs extracted by the CEEMDAN for the structure before (healthy) and after damages of 20%, 50%, 80%, and 100%, respectively. The damage of 100% means the removal of the diagonal element from the damage location. The first three natural frequencies of the bridge are 19.53, 40.16, and 60.55 Hz, and the corresponding natural period times T_1, T_2, and T_3 are 0.051, 0.025, and 0.016 s, respectively. Hence, considering 1.0 s of the acceleration response of the bridge (about $20T_1$) is a reasonable time window to analyze in this study.

In Figure 5.2b, the damage spikes are observed in the time-history behavior of the IMFs for the damaged cases of the structure as to be expected. By comparing the first three IMFs, it is observed that the intensity of the

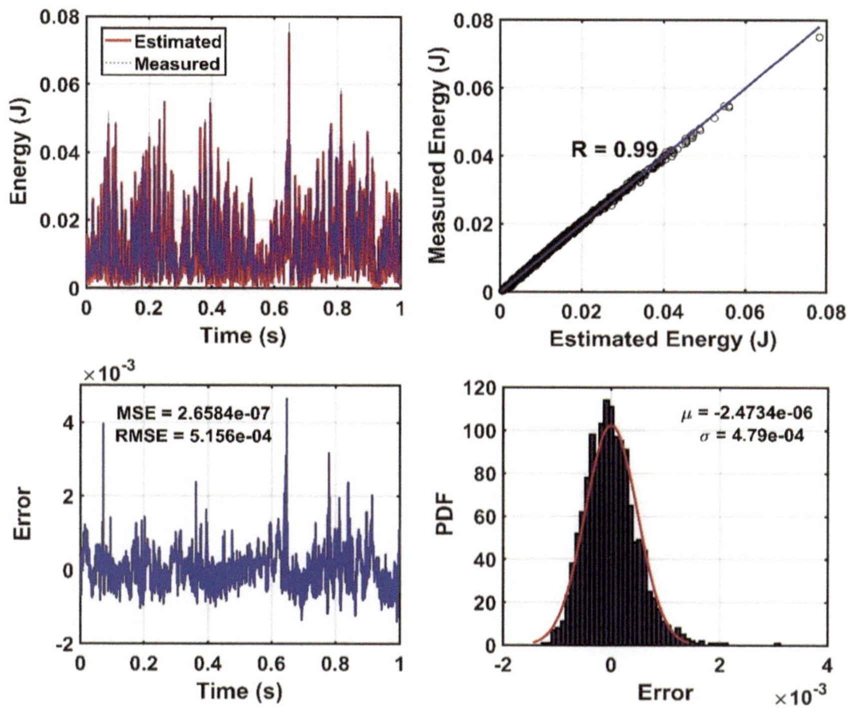

Figure 5.3 Evaluation of the accuracy performance of training the ANN based on the energy parameter of sensor-10 for the healthy state of the truss. (where *R*, *μ*, *σ*, MSE and RMSE denotes the value of regression, Mean, standard division, mean square error, and root mean square error, respectively).

spikes of IMFs becomes more pronounced as the level of damage increases. Furthermore, the root-mean-square (RMS) values which represent the total behavior of the IMFs increase by enhancing the damage level. Once the signal is decomposed by the CEEMDAN, the Hilbert Transform (HT) is applied to the IMFs captured from the sensors before and after damage to extract four key features including the energy, IA, unwrapped phase, and IF. Then, the obtained IMFs and the features are used to train the ANN.

Before the use of the proposed model, the reliability of the CEEMDAN-HT-ANN model is examined through the correlation and root-mean-error analyses between the measured energy of the IMFs and the estimated (output) energy by the proposed model in the healthy state of the truss bridge. Figure 5.3 demonstrates the accuracy of the training and testing process of the ANN. In addition, the comparison between the measured energy of the IMFs and the estimated energy by the proposed model for the 100% damage state of the truss is shown in Figure 5.4. It is found that the CEEMDAN-HT-ANN model can efficiently and accurately perform in estimating the results of both healthy and damaged states of the truss.

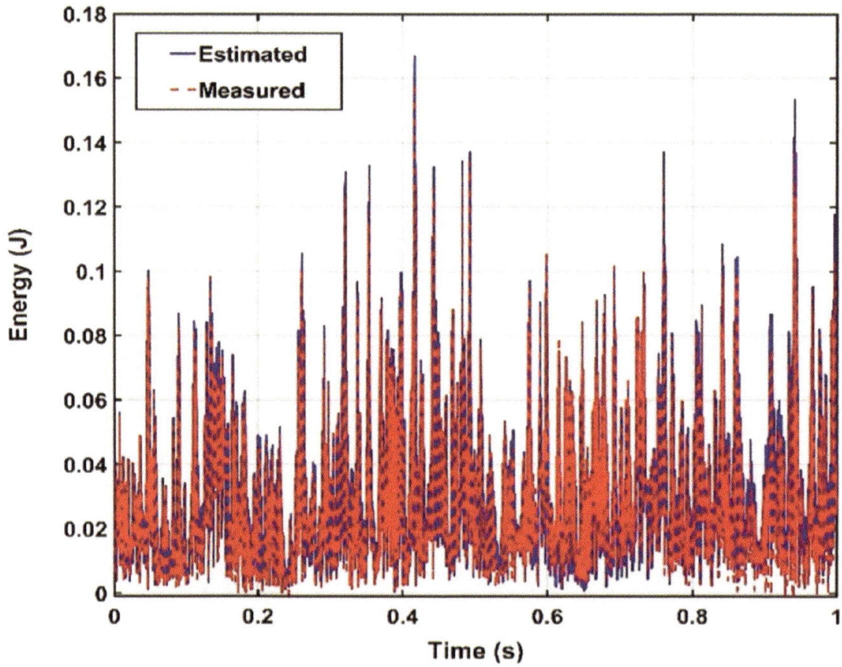

Figure 5.4 Comparison of the IMFs' energy of sensor-10 between the outputs of the trained model (i.e., estimated) and the measured data for 100% damage state of the truss.

Due to the successful performance of the ANN, all the IMFs of 13 sensors in 20%, 50%, 80%, and 100% damaged states of the structure are used to examine the ANN model. The outputs of the CEEMDAN-HT-ANN model based on the four features (i.e., energy, IA, unwrapped phase, and IF) are compared to show the capability of the proposed model in detecting the presence and severity of the damage in the truss bridge. Figure 5.5 illustrates the estimated energy using the CEEMDAN-HT-ANN model for the acceleration response of sensor-10 for healthy and different damaged states of the structure. The increase of energy power is observed in proport ion to the increase of the damage level. As the second extracted feature, the estimated *IA* by the proposed model also increases with the increase of the damage level as shown in Figure 5.6. Besides, the estimated unwrapped phases of the IMFs using the CEEMDAN-HT-ANN model before and after damage states of the truss are depicted in Figure 5.7. It is found that the unwrapped phase of the IMFs decreases and their deviation increases relative to the healthy unwrapped phase with increasing the level of damage. Consequently, the outputs of the proposed model based on four extracted key features demonstrate the significant positive influences of applying the CEEMDAN and Hilbert transform combined with the ANN in detecting and classifying the damage severity.

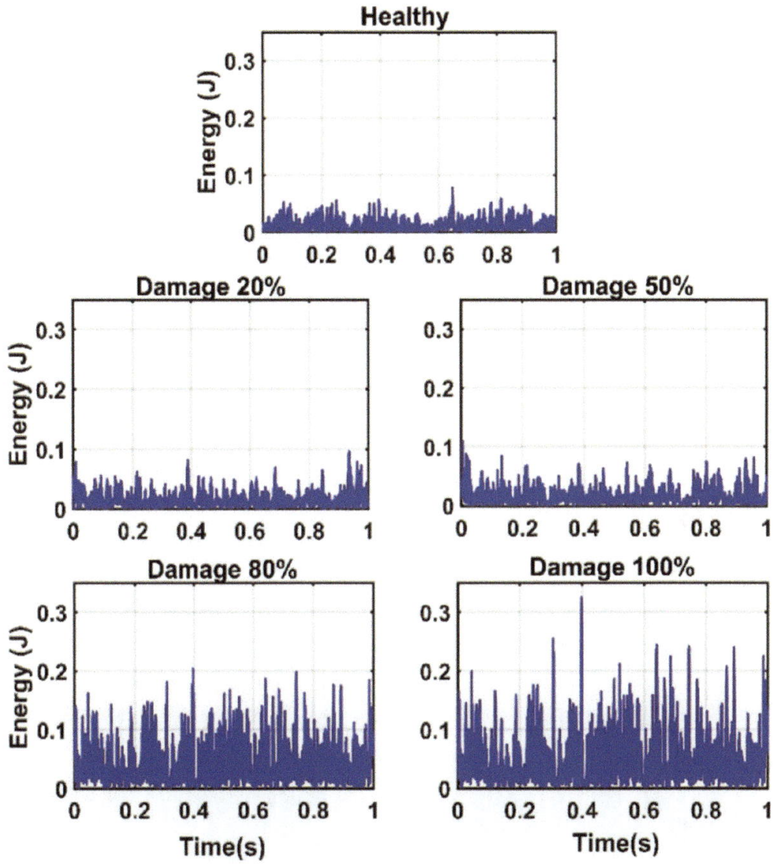

Figure 5.5 The energy of IMFs of the sensor-10 estimated by the proposed model for the healthy state and different damage levels of the truss.

The efficiency of the proposed model is more explored by evaluating the results of damage indices based on four features as given in Table 5.1. The results demonstrate the significant enhancement in the values of the damage indices with increasing the damage level. Furthermore, It is obtained that the damage index values based on the energy feature captured significantly higher values than those from IA and unwrapped phase parameters.

The IF is considered as the fourth feature to detect the severity of the damage. Generally, when a structure experiences nonlinear behaviors, changing of structural stiffness causes the change of frequency responses which can consequently imply the existence of damage. The three-dimensional spectrograms of the IMFs are presented in Figure 5.8a–d. These figures show the high-resolution spectrums of the HHT in which the amplitude of the signal is presented in a time-frequency domain. The sampling frequency of the spectrums is 500 Hz while the length of overlapping windows is 100 samples

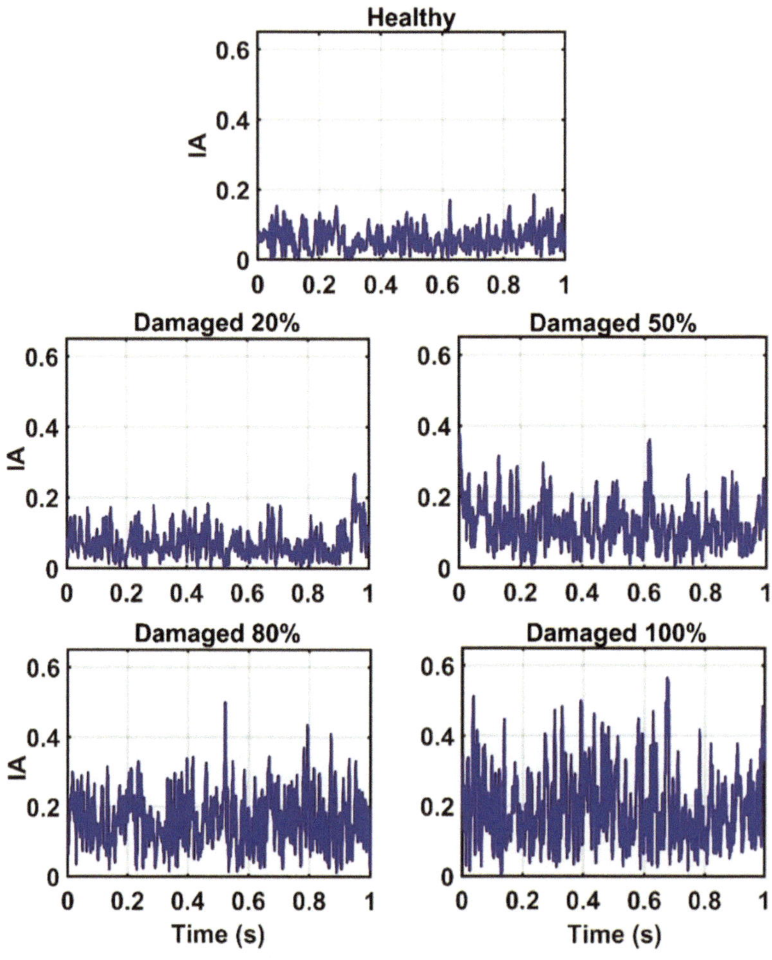

Figure 5.6 The estimated IA of the sensor-10 by the proposed model for the healthy state and different damage levels of the truss.

based on STFT. In Figure 5.8, it is observed that when the level of damage increases, the intensity (power) of *IF* increases in the low-frequency range especially for 80% and 100% damage levels compared to the healthy state.

5.3.3 Detection of damage location using CEEMDAN-HT-ANN model

To detect the location of damage, the vibration responses of the structure in the healthy and damaged states are captured from four different sensors 10, 9, 4, and 1 which represent the nearest, the second near, the far, and the

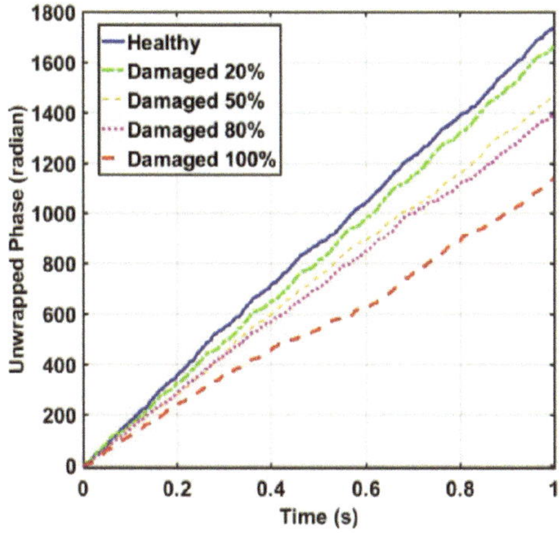

Figure 5.7 Comparison of the IMFs' unwrapped phase of acceleration responses of sensor-10 estimated by the proposed model for the healthy and different damage levels of the truss.

Table 5.1 Damage index results based on the energy, IA, and unwrapped phase of IMFs for the different damage levels

	Damage 20%	*Damage 50%*	*Damage 80%*	*Damage 100%*
IMFs' feature	*DI (%)*	*DI (%)*	*DI (%)*	*DI (%)*
Energy Ikkljlklkl	23.43	32.72	86.47	94.24
Amplitude	21.46	25.88	57.39	62.04
Unwrapped phase	6.52	17.13	20.11	41.80

farthest sensors from the location of the damaged element, respectively. In this section, the damage scenario is implemented by removing the diagonal element (i.e., damage 100%). Thereafter, the collected vibrations from the aforementioned sensors are decomposed by the CEEMDAN and the four features including energy, IA, unwrapped phase, and IF are extracted from the IMFs. Then, the outputs of the CEEMDAN-HT-ANN model based on these features are quantified to assess the performance of the proposed model in detecting the location of the damage in the truss bridge. Figures 5.9–5.11 illustrate the comparison of energy, *IA*, and unwrapped outputs from the proposed model between those obtained for the healthy and damaged states of the truss at different sensor locations relative to the damaged element. It is observed that the intensities of the energy and *IA* spikes obtained from the sensors located at near distances from the damaged element are greater than those located at far distances. That is, the estimated energy and *IA* as the outputs of the CEEMDAN-HT-ANN model increase with decreasing the distance of the sensor location from the damaged element that demonstrate the ability of these parameters in detecting the location of the damage.

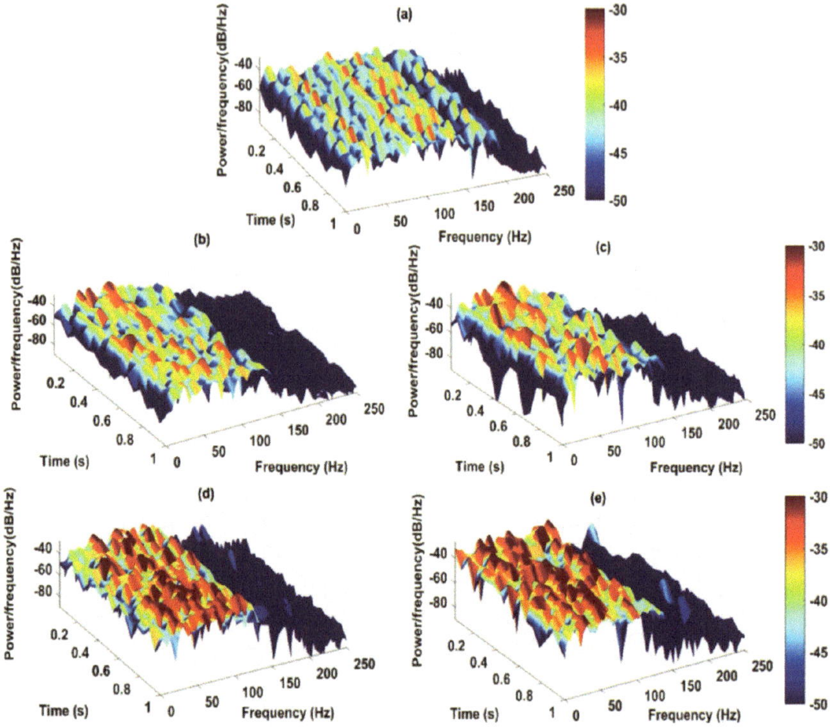

Figure 5.8 Spectrograms output of the IMFs of sensor-10 under five states of the structure including; (a) healthy, (b) 20% damage, (c) 50% damage, (d) 80% damage, and (e) 100% damage by the CEEMDAN-HT-ANN model.

Besides, the estimated unwrapped phase as one the output of the CEEMDAN-HT-ANN model is assessed to detect the damage location. In Figure 5.11, it is obviously observed that the value of the unwrapped phase decreases by reducing the distance between the specified sensors and the damaged element and their deviations increase relative to the healthy unwrapped phase. In addition, the damage indices based on the features are given in Table 5.2. It is found that although all features are successfully able to locate the damage, the energy-based damage index captures a more efficient approach compared to those based on the IA and unwrapped phase parameters.

The IF is considered as the fourth feature to detect the location of damage through the three-dimensional spectrograms. This approach presents the IF of IMFs in a time-frequency domain. The power of *IF* increases with decreasing the distance of sensors from the damaged element as shown in Figures 5.12a–d. Upon careful observation of these figures, the spectrogram of the nearest and second-near sensors shows the increase of power density in the low-frequency spectra compared to those from far and

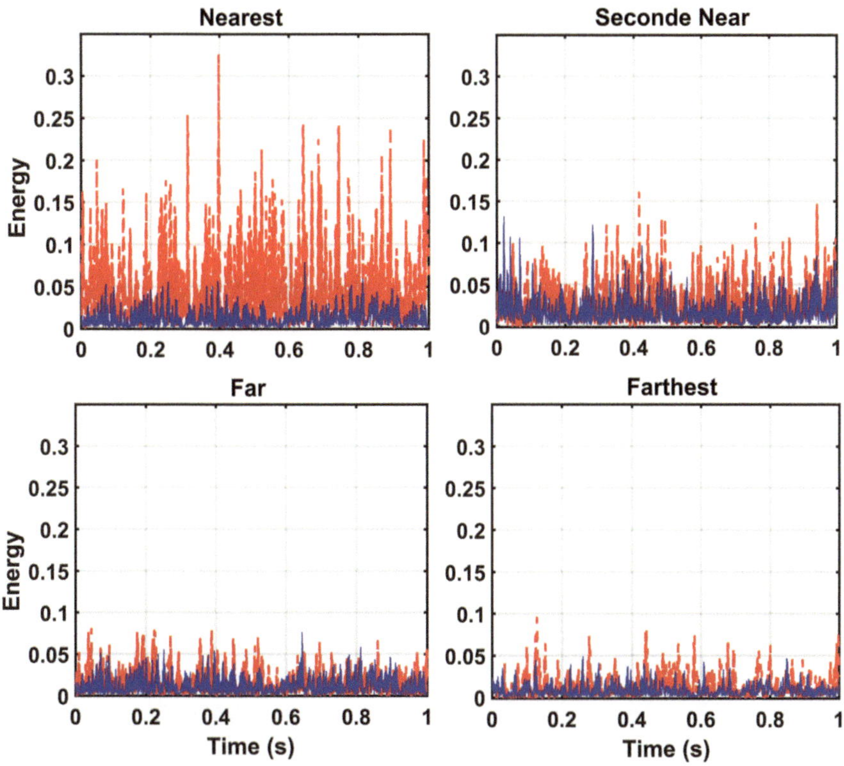

Figure 5.9 The energy output of the IMFs for different sensor locations from the damaged element by the proposed model.

farthest sensors. That is, there is a decrease in the power of the frequency as the sensor location is moved farther with respect to the damage location. Generally, the extracted features evaluated in this section were successfully able to locate the damage and demonstrate the performance of the proposed methodology both quantitatively and qualitatively in classifying the damage severity and locating the damage.

5.4 HYBRID APPLICATION OF EWT AND ANN TECHNIQUES

The procedure of the proposed damage diagnosis approach named EWT-ANN is introduced in this section. In the first step of this method, the vibration signals of the steel truss bridge in healthy and four damaged states are recorded by 13 sensors when exposed to the band-limited white noise and impact excitations. Then, the EWT is adopted to decompose the original signal into several modulated signal modes (here, 12 modes). To train the

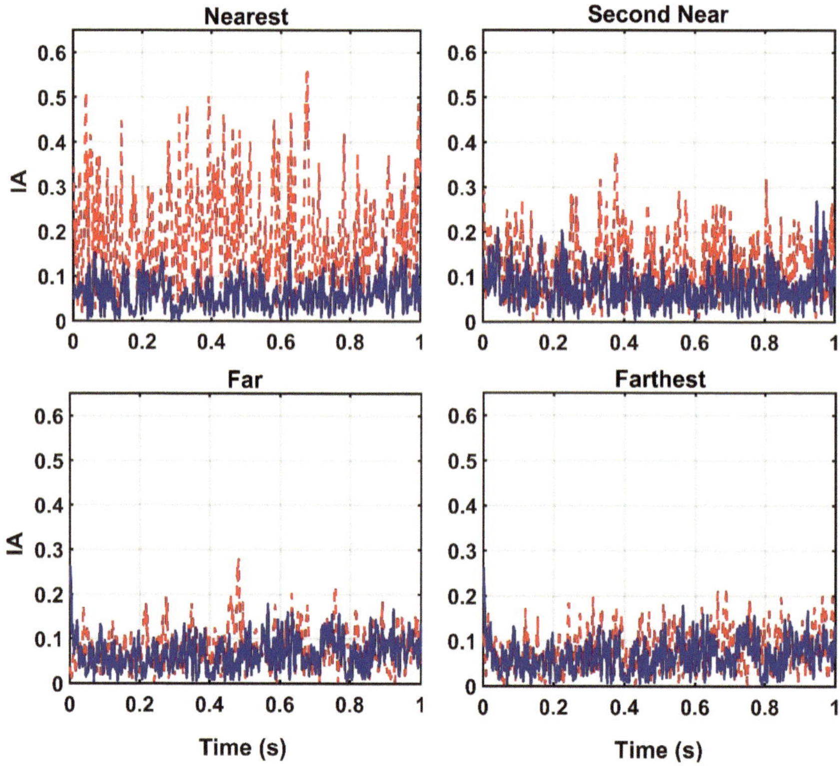

Figure 5.10 The IA output of the IMFs for different locations from the damaged element by the proposed model.

EWT-ANN model, the signal components from the healthy state of the bridge are defined as the inputs to the ANN. Afterwards, the statistical time-domain features including RMS, shape factor, kurtosis, and entropy are automatically extracted inside the trained model from the generated signal modes to define them as the outputs to the ANN. Therefore, the ANN trains relationships between the signal components and the extracted features as the input and output layers, respectively, for the healthy state of the bridge and the outputs of the trained EWT-ANN procedure for the healthy truss is considered as the target. Thereafter, to operate the automated EWT-ANN damage detection model, the signal modes resulted from all 13 sensors for different damage scenarios of the truss are tested using the trained model based on the healthy state of the bridge to automatically extract the signal features and calculate the damage indices inside the EWT-ANN model. That is, for the vibration signals recorded by each sensor, only one ANN is trained to conduct the proposed EWT-ANN approach. Consequently, the EWT-ANN outputs resulted from the damaged states compared to the trained targets (i.e., the outputs obtained from the healthy

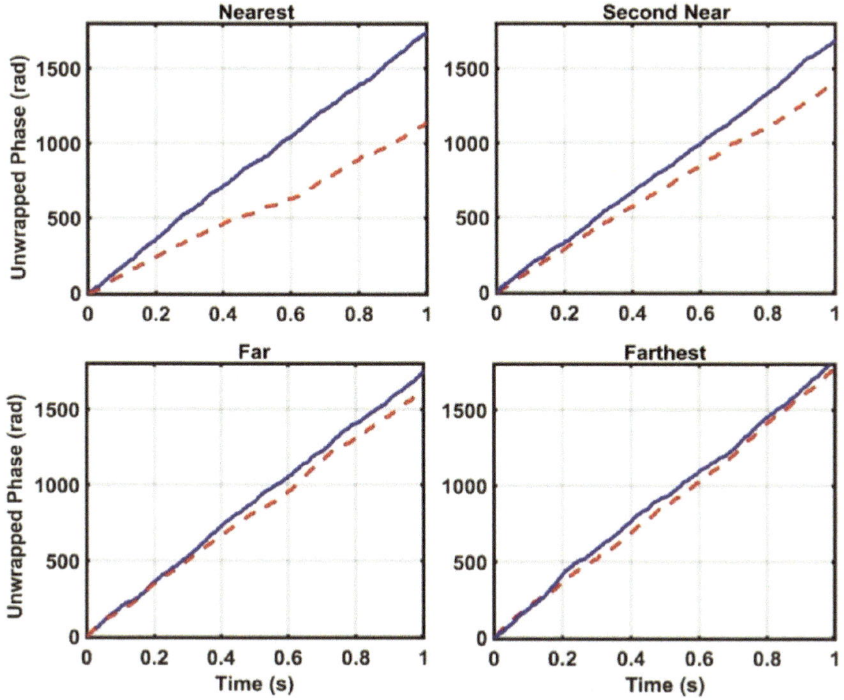

Figure 5.11 Comparison of the IMFs' unwrapped phase outputs of the acceleration responses by the proposed model for different sensor locations from the damaged element.

Table 5.2 Damage index results based on the IMFs' features for different sensor locations from the damaged element

	Nearest	*Second near*	*Far*	*Farthest*
IMFs' feature	*DI (%)*	*DI (%)*	*DI (%)*	*DI (%)*
Energy	94.24	35.84	24.92	17.34
IA	62.04	51.36	31.35	16.45
Unwrapped Phase	41.80	18.90	07.55	05.10

truss) are evaluated to qualitatively and quantitatively identify the presence, location, and severity of the damage. Figure 5.13 illustrates the framework flowchart of the proposed approach trained based on the signal modes and features of the bridge in a healthy state. Further descriptions of this flowchart are given in Section 5.4 [340].

For the truss bridge case study in this study, if ANN does not be used, the feature extraction (five signal features) and damage index calculation

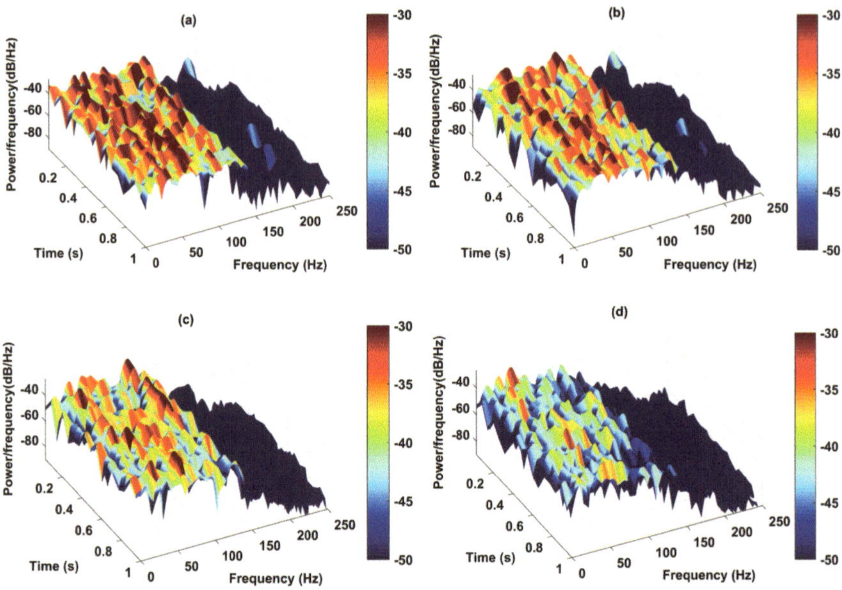

Figure 5.12 Spectrograms output of the IMFs for different locations of (a) the nearest, (b) second-near, (c) far, and (d) the farthest sensors from the 100% damaged element through the proposed model.

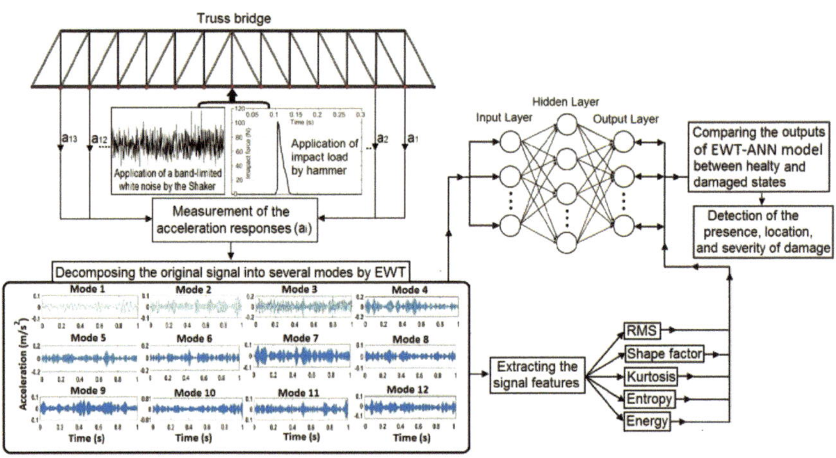

Figure 5.13 The framework of the proposed EWT-ANN damage detection approach trained for the healthy state of the bridge.

processes should be manually repeated for $2\times5\times5\times13=650$ times for the vibration responses of all (13) sensors in healthy and four damage scenarios (totally five scenarios) and five features including RMS, shape factor, kurtosis, and entropy from the signal components decomposed by EWT under two types of excitations which would take a too long time. Therefore, the use of ANN can significantly reduce costs and computation time.

5.4.1 Feature extraction process

Despite the extensive use of statistical time-domain features such as root mean square (RMS), standard deviation, variance, kurtosis, skewness, shape factor, crest factor, and entropy in fault diagnosis of mechanical systems [494], they have been rarely used in the damage detection process of complex civil structures. The first three features are commonly employed to recognize the differences between the two vibration signals. For non-stationary signals, skewness, kurtosis, and entropy features which evaluate the PDF of signals, are more effective. Also, shape and crest factors which are operated based on the changes in the information of the mean, RMS, and standard deviation of the signal, can obtain the non-dimensional information of the analyzed signal.

To identify the differences among the vibration responses of the different states of the structure, in addition to the energy feature of the signal components, four statistical time-domain features including RMS, shape factor, kurtosis, and entropy are extracted. According to the literature, with any abrupt changes in the intensity of the vibration signal in various case studies, the values of RMS and energy features of the signal increase [479,495]. Therefore, the RMS and energy could be proposed to establish an effective indicator in identifying the damage.

The energy (E) of a given signal, x (t), decomposed by EWT can be calculated from the sum of the square of the signal data as follow:

$$E = \int_{0}^{t} x^2(t)\,dt \tag{5.7}$$

Among the statistical features, the RMS is usually used to quantify the difference values between two vibrational signals. According to the literature [477], it has been proven that the value of RMS increases in proportion to the progress of a system fault or damage. The RMS of a given signal x_i (i.e., the sampled vibration signal) with N number of data points can be calculated as follows:

$$\text{RMS} = \sqrt{\frac{1}{N}\sum_{i=1}^{N} x_i^2} \tag{5.8}$$

As a non-dimensional statistical feature, the shape factor (SF) of a signal can be defined from the ratio of RMS to mean of the sampled signal data as follows:

$$\text{SF} = \frac{\sqrt{\frac{1}{N}\sum_{i=1}^{N} x_i^2}}{\frac{1}{N}\sum_{i=1}^{N} |x_i|} \tag{5.9}$$

Since the kurtosis (Ku) is calculated using the peak of signal PDF as given in Equation (5.10), its value increases with any changes that occur in the PDF of vibration responses due to the appearance of damage. Therefore, this feature would be more effective in analyzing the non-stationary signals.

$$Ku = \frac{\sum_{i=1}^{N}(x_i - m)^4}{(N-1)\sigma^4} \tag{5.10}$$

where m and σ denote the mean and standard deviation of the signal.

In addition, various damage detection and SHM-based studies demonstrated the increment of entropy value with increasing the complexity of vibration response of structures that indicate the existence of damage [415,493]. The entropy, $e\ (p)$, is known as another statistical feature that is dependent on the PDF of a vibration signal. This feature is calculated based on the histogram of the PDF and represents the uncertainty and the randomness of a sampled signal given as follows:

$$e(p) = -\sum_{i=1}^{N} p(x_i)\log_2 p(x_i) \tag{5.11}$$

where $p\ (x_i)$ denotes the normalized histogram of the sampled signal x_i.

5.4.2 Limitations of the proposed methodology

Similar to the damage scenarios considered in the previous section, four damage scenarios are also considered for the truss bridge in this study. Three of these scenarios are implemented by replacing a diagonal element located between the locations of sensors 10 and 11 as illustrated in Figure 5.14. The severities of three damaged states of the bridge are accomplished by reducing 35%, 60%, and 83% of the cross-sectional stiffness of the specified element as shown in Figure 5.14. The reductions of the cross-sectional moment of inertia of these damaged elements in percentage terms are; (i) 35% in which the outer and inner diameters are 16 and 10 mm, respectively, (ii) 60% in which the outer and inner diameters are 14 and 8 mm, respectively, (iii) 83% in which the diameter of the bar element is 11 mm. Besides, the 100% damaged state of the bridge is designed by removing the specified diagonal element. The axial stiffness reductions of these damaged elements are 14%, 27%, and 33%, respectively. It is noteworthy that excluding the aforementioned diagonal element, other structural characteristics of the bridge and the loading excitations are kept constant for all the healthy and damaged scenarios.

Similar to the white noise used in Section 4.2, a band-limited white noise excitation with a bandwidth of 100 Hz as shown in Figure 5.15a (for a partial period of 1.0 s) is applied to the truss. In addition, the impact excitations are applied to the bridge using a typical hand-held impact hammer.

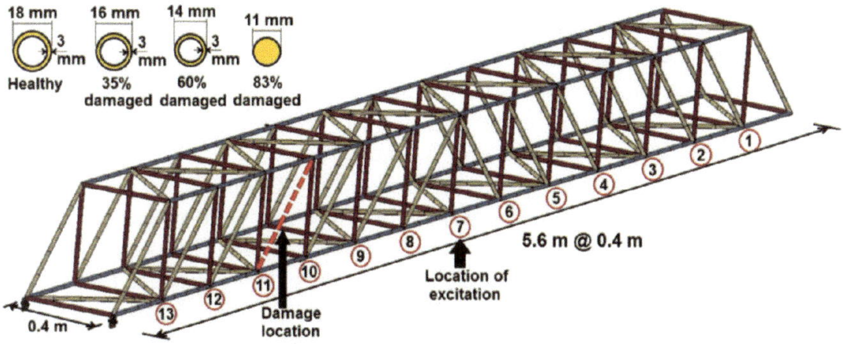

Figure 5.14 A schematic presentation of the truss bridge model by showing the sensor locations and the cross-sectional dimensions of truss elements in healthy and different damaged states.

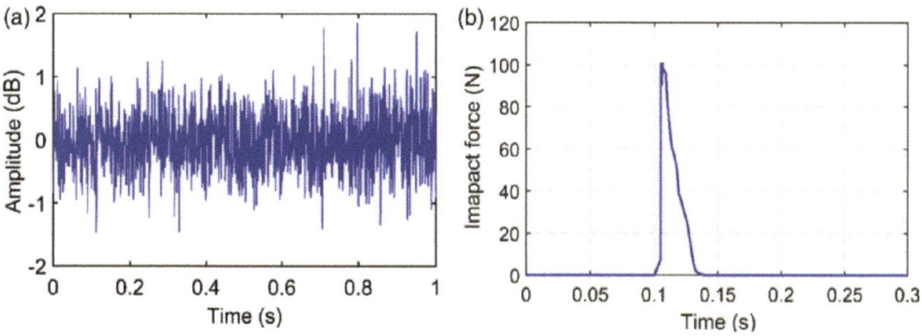

Figure 5.15 Excitations applied to the truss bridge model; (a) a band-limited white noise by the shaker, (b) an impact impulse by a typical hand-held impact hammer.

The frequency content of the input impact force and consequently the input spectrum is extremely dependent on the stiffness, mass, and velocity of the struck hammer. In this study, an impact impulse with a peak input force of 100 N, and a duration of about 50 ms is applied to the bridge as shown in Figure 5.15b. This impact load represents an impulsive loading type since the duration of the impact load equal to 35 ms is less than the first natural period of the truss bridge in its healthy state which is 51 ms (with the corresponding first natural frequency of 19.53 Hz) [401].

There exist several limitations in the implementation of the experimental tests that can be listed as follows:

- Sensor arrangement: Due to the limitations of sensor arrangements on the truss elements of the laboratory-scaled truss bridge (because of the lower-dimensional surface of elements), a uniform distribution was selected in which the sensors were attached to the joints rather than truss elements.

- Manufacturing of the truss elements for the damage scenarios: Due to existing limitations for fabricating steel truss elements, the damage scenarios were performed by using only the damaged elements with producible diameters in the factory (only tubes with a thickness of 3 mm) as considered in this study with percentage reductions of the cross-section area of 35%, 60%, and 83%.
- Location of damaged element: To avoid redundancy and lengthening this study, the damage scenarios of the truss were implemented by assuming the damaged states for only one diagonal truss element.
- Frequency of excitations: To eliminate the effects of other factors on the results and increase the distinguishability and accuracy of the assessment procedure, the same excitations are applied to the truss bridge in healthy and damaged states to carry out all the scenarios in the same loading conditions
- Loading application: Owing to the limitations of shaker facilities, both white noise and impact excitations were applied to the mid-span of the truss bridge using a single shaker.

5.4.3 Detecting the presence, severity, and location of the damage under white noise excitation

This section aims to identify the damage existence and intensity for various damage scenarios of the bridge model when subjected to a white noise excitation with a bandwidth of 100 Hz as shown in Figure 5.15a. White noise is a random signal having equal intensity at different frequencies which can be resulted from various environmental excitations such as traffic, wind, walking in the real world.

Figure 5.16a shows the 1.0 s acceleration response with a sampling frequency of 2000 points per second from sensor-10 as an example checkpoint in the healthy bridge. As discussed in the previous study [496], selecting 1.0 s of the bridge acceleration is reasonably sufficient since it is about 20 times of the first natural period of the bridge with 0.051 s and the first natural frequency of the bridge is 19.53 Hz as shown in Figure 5.16b which illustrates the FFT spectrum of the acceleration signal. Also, the second and third natural frequencies of the bridge model are 40.16, and 60.55 Hz as shown in Figure 5.16b. Moreover, the mode components of the vibration response of sensor-10 decomposed by EWT are illustrated in Figure 5.16c. The output of EWT is composed of the filtering with the scaling function and the N number of wavelets. The EWT automatically estimates the number of modes for different original signals based on the number of wavelets which represents the number of spectrum segments with separate boundaries. The number of these wavelets is detected by EWT using the number of most important maxima in the magnitude of the FT of the signal and their detected bands. The procedure of estimating the number of modes by EWT can be found in the literature [475]. The segmented boundaries and 11 local

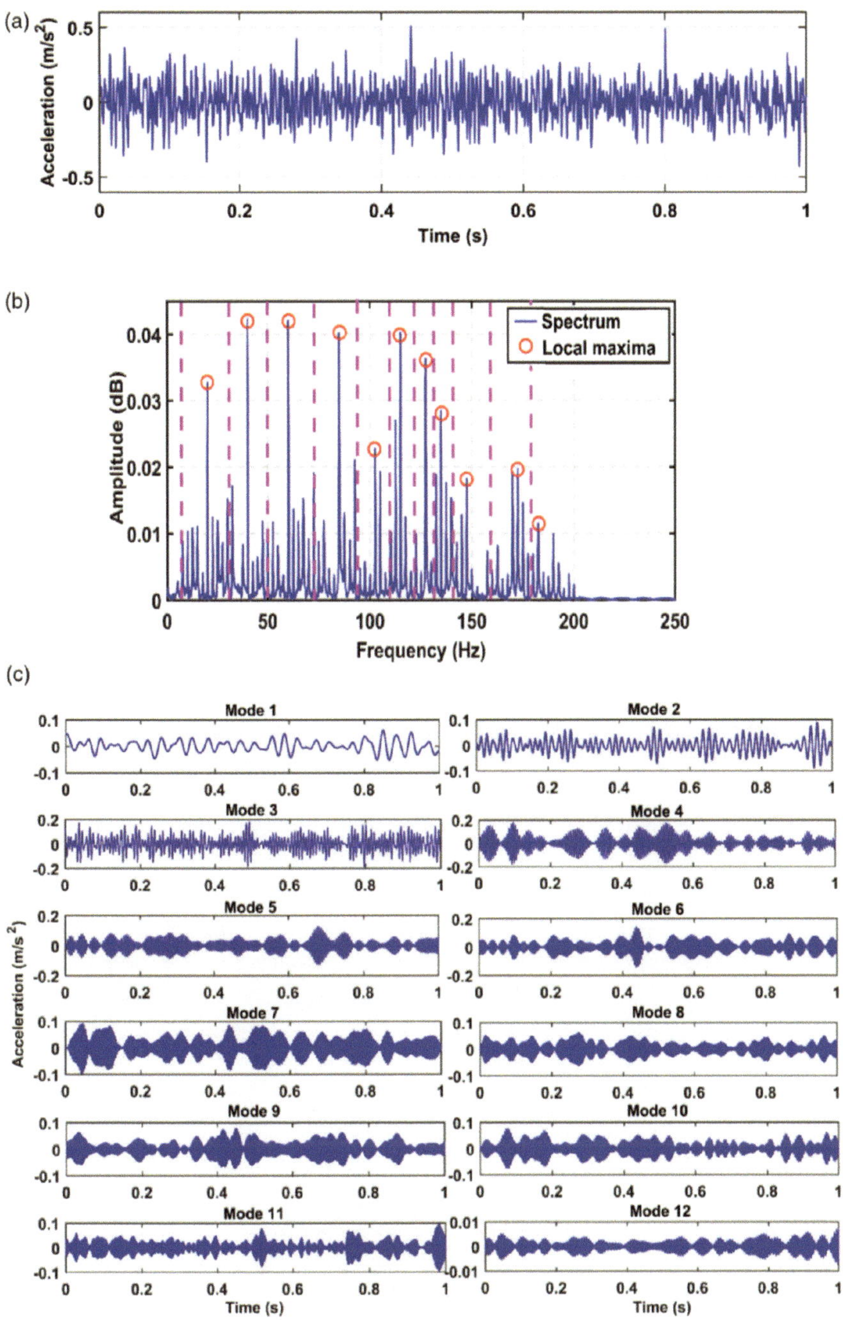

Figure 5.16 (a) Acceleration signal recorded by sensor-10 for the healthy bridge under white noise, (b) the detected boundaries using the local maxima of the signal spectrum by EWT, (c) corresponding signal components decomposed by EWT.

maxima determined by pink dashed lines and Red circles in the FFT spectrum of the original acceleration response (in Figure 5.16a) are illustrated in Figure 5.16c. In the estimation of the number of modes by the EWT, this technique overestimates the number of modes relative to the number of local maxima determined to keep all these local maxima as discussed in the work done by Gilles [475]. Therefore, in this study, the number of modes of the original vibration signal estimated by EWT is 12 as illustrated in Figure 5.16c.

As discussed in Section 5.3, to classify the damage severity, the proposed damage detection approach is applied to the acceleration signals recorded by sensor-10 for various damage scenarios of the bridge. The three-dimensional plots of the 12 mode components of the vibration signal decomposed by the EWT are illustrated in Figure 5.17a–e for the bridge in healthy and damaged states with the severity of 35%, 60%, 83%, and 100%, respectively.

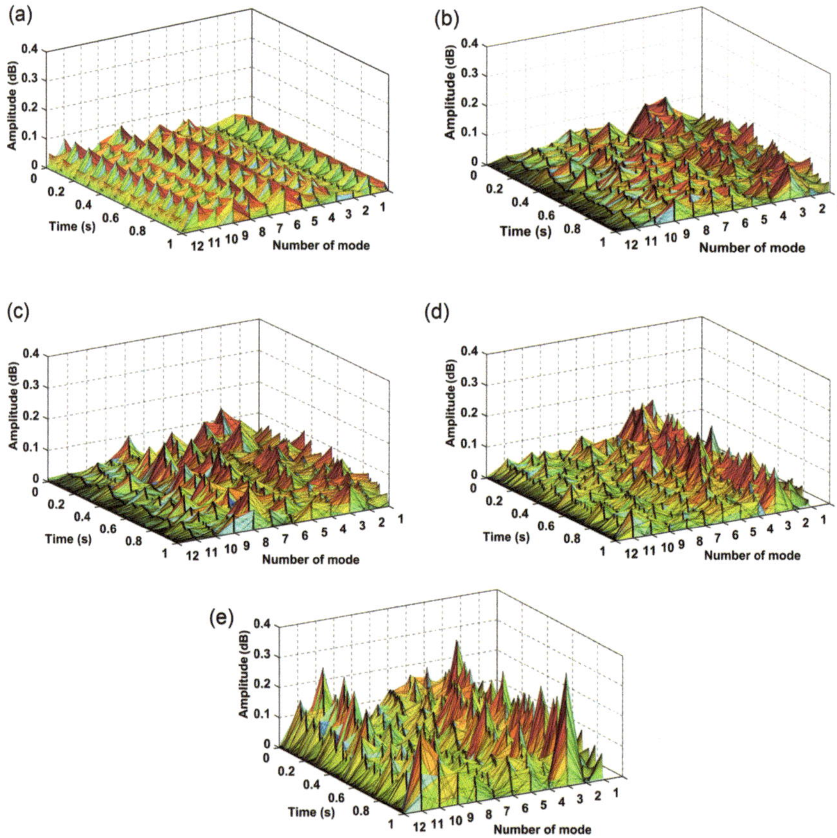

Figure 5.17 Mode components of the vibration signal of sensor-10 under white noise obtained by EWT for the various states of the bridge including; (a) healthy, (b) 35% damaged, (c) 60% damaged, (d) 83% damaged, and (e) 100% damaged.

It is seen that the intensity of the spikes and the amplitude of the modes increase with enhancing the damage intensity. Hence, the advantage of presenting these three-dimensional (3D) plots is that several time-history plots and their changes such enhancement trend of the amplitudes can be observed at the same time. For further assessments, the time-domain features of these signal modes including energy, RMS, shape factor, kurtosis, and entropy are extracted from the mean of the signal modes, analyzed from the vibration responses of the 13 accelerometer sensors that are uniformly distributed on the joints of the lower chord of the bridge model. All the mode components and obtained features before implementing the damage scenarios (i.e., the healthy state) are used to train the ANN as the input and output layers, respectively.

The reliability of the proposed EWT-ANN model as an automated damage detection approach is assessed using the correlation analysis of the measured energy of the signal components in comparison with the estimated energy using the proposed approach for the healthy bridge. Figure 5.18 illustrates the validity of the training and testing processes. By observing the analysis results of the regression (R), Mean (μ), standard division (σ), mean square error (MSE), and root mean square error and (RMSE), the capability of the EWT-ANN approach in estimating accurate results was

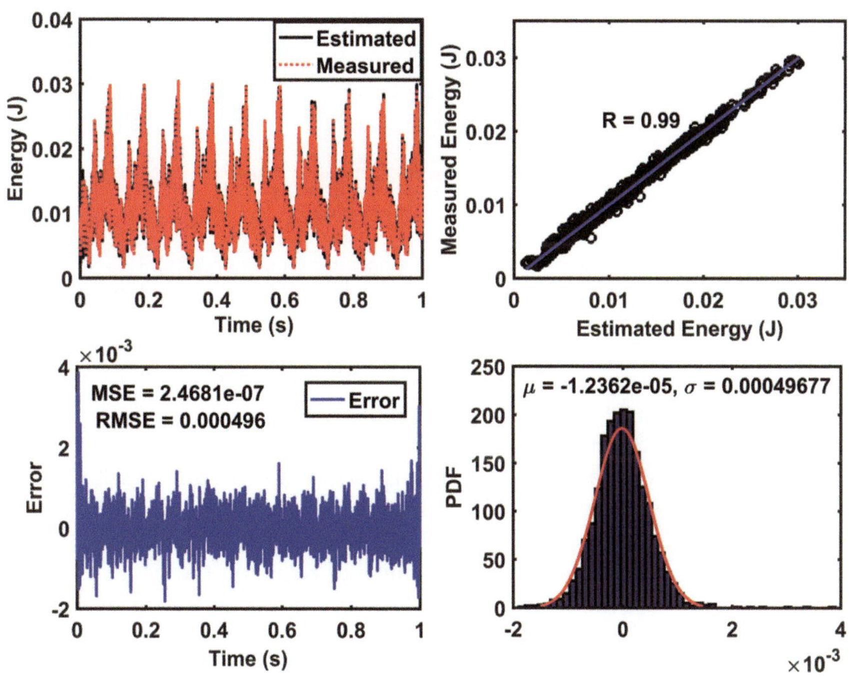

Figure 5.18 Assessing the validity of the ANN in training the energy feature of sensor-10 for the healthy bridge.

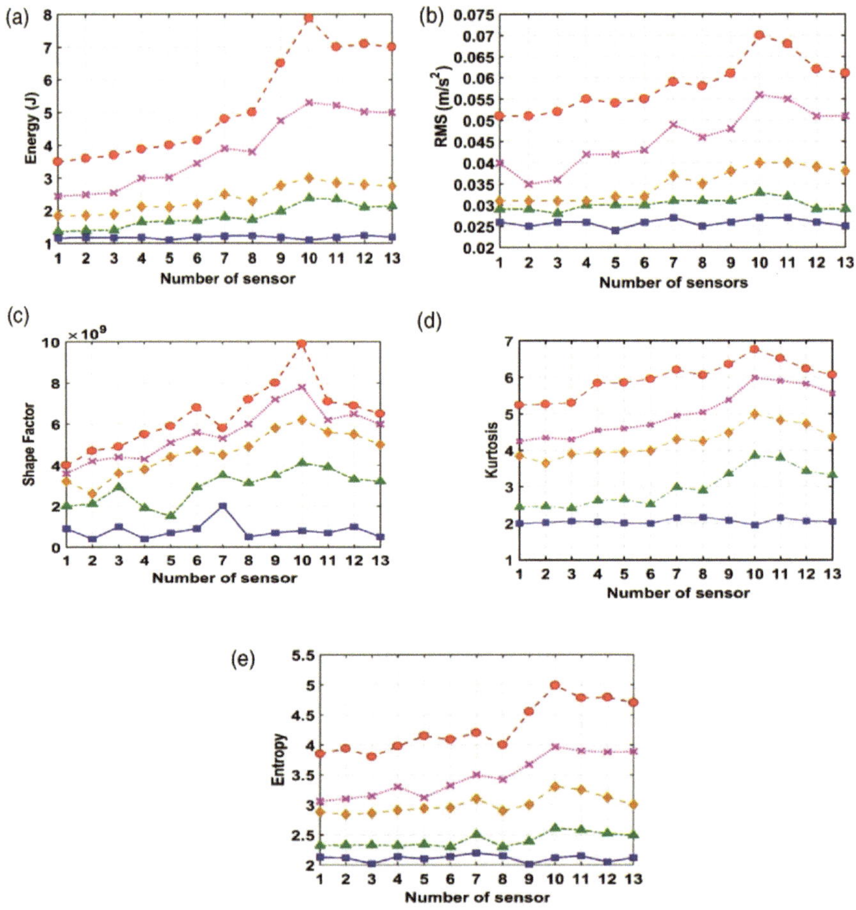

Figure 5.19 The estimated time-domain features including; (a) energy, (b) RMS, (c) shape factor, (d) kurtosis, and (e) entropy, by the EWT-ANN model from 13 sensors under a white-noise excitation.

concluded. Therefore, it can be considered that the output of the trained model is reliable to be examined by the damaged states.

Following the accurate process of the ANN as demonstrated above, the mode components of the vibration signals acquired from the 13 sensors for different damage states of the bridge with 35%, 60%, 83%, and 100% severities, are adopted to assess the EWT-ANN model. The comparison of output results from the EWT-ANN method computed based on the signal features for various states of the bridge is illustrated in Figure 5.19 to demonstrate the localization and quantification of the damage severity. It is seen that the magnitudes of the damage indices based on the five signal features increase as the damage severity is enhanced. Also, it is realized that the damage index curves have more concentrated curvature with higher values around the damage location at the sensor-10. Therefore, these figures demonstrate the performances of the damage indices calculated based on EWT-ANN in identifying the presence, location, and severity of the damage altogether.

As a comparative study, the performance of the EMD technique in combination with ANN (i.e., EMD-ANN) in identifying the severity and location of the damage in the truss is evaluated and the results are illustrated in Figures 5.20a to the based on the five features. It is noteworthy that

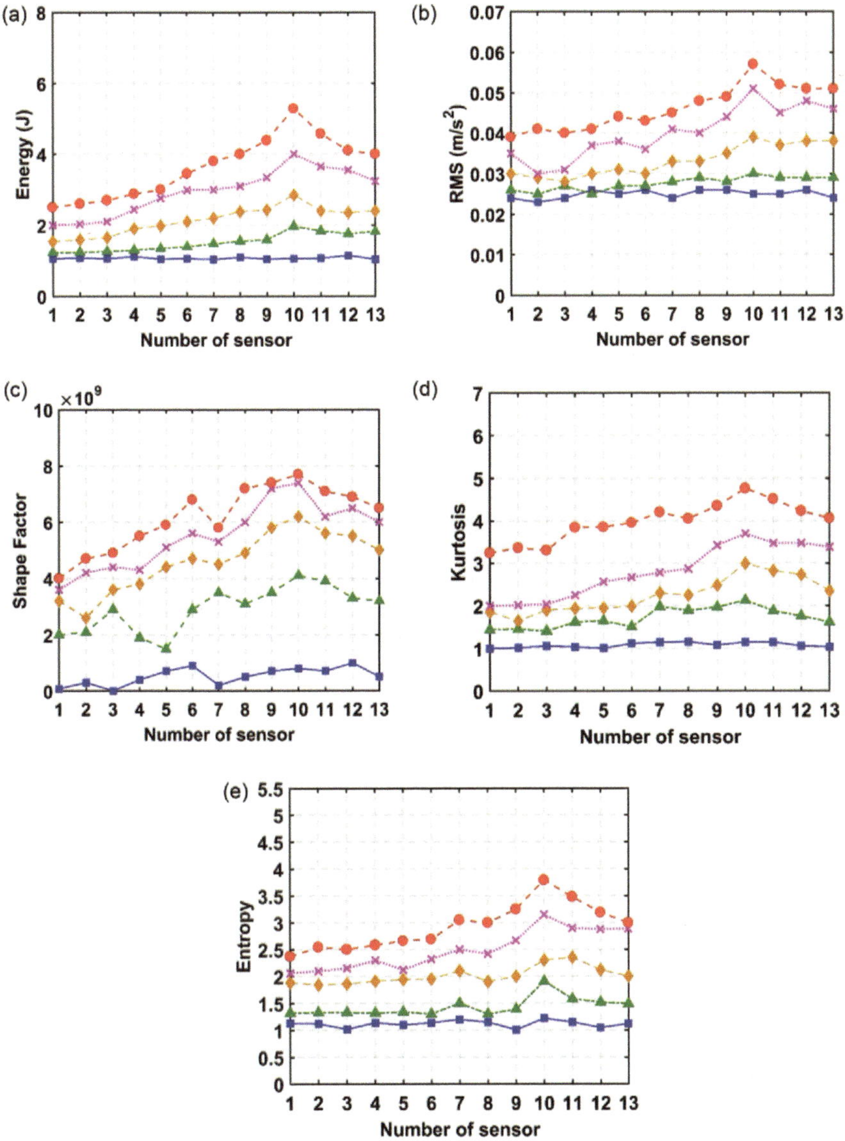

Figure 5.20 The estimated time-domain features including; (a) energy, (b) RMS, (c) shape factor, (d) kurtosis, and (e) entropy, by the EMD-ANN model from 13 sensors under a white-noise excitation.

EMD-ANN approach is implemented using a similar methodology to that of the EWT-ANN approach as introduced in Section 2.3, with the difference that EMD is employed to decompose the acceleration signals into a series of IMFs [283] which are used as the inputs of the ANN. By comparing the results given in Figures 5.19a–e and 5.20a–e, it is found that the EWT-ANN approach demonstrates more accurate and efficient performance in identifying the damage severity and its location based on the signal features compared to EMD-ANN, even for the damage severities smaller than 60%. As a result of a 100% damage scenario, the EWT-ANN method results in higher magnitudes for the features including 32% of energy, 17% of RMS, 20% of shape factor, 15% of kurtosis, and 24% of entropy in comparison with those of EMD-ANN around the damage location. Therefore, the superiority of the EWT-ANN method in damage identification and localization can be concluded in comparison with the performance of the EMD-ANN method.

Moreover, the spectrogram of the mean of mode components is considered as a feature in the frequency-time domain to determine the damage level. The reason for adopting the spectrograms in this study is to demonstrate both amplitude and frequency characteristics of the analyzed data in a time-frequency-energy domain with high resolution. There exist many research works in the literature utilizing spectrograms to detect the damage in different systems. According to the literature [401,496–498], it has been proven that the energy of spectrums and the vibration magnitude increase at lower frequency ranges with the increase of the damage level due to the stiffness loss. In other words, the damaged structure exhibits different degrees of frequency lowering which represents the presence of structural nonlinearity in the system caused by stiffness loss. This frequency reduction behaviour is due to the decrease of the square root of the stiffness of the structure in the presence of the damage. It is noteworthy that all the environmental conditions in the laboratory including the temperature were similar during the implementation of all healthy and damage scenarios on the truss bridge. Therefore, these frequency changes could not be resulted in due to environmental causes. The high-resolution spectrums of the mean of mode components are presented using 3D time-frequency plots in Figures 5.11a–e. These spectrums have a sampling frequency of 500 Hz and a length of 100 samples for overlapping windows. It is seen that power of spectrums increases in the lower frequency ranges with enhancing the damage intensity. It is noteworthy that the frequency power is increased between the ranges 100 and 120 Hz for the healthy bridge (see Figure 5.21a). However, the frequency power is enhanced in the lower ranges between the ranges 10 and 15 Hz with the increase of the damage severity especially for 83% and 100% damages as shown in Figure 5.21d and e. This shift to lower values in the peaks of the frequency power as the damage level increase is due to the reduction in stiffness and the concomitant dominant natural frequency of the bridge.

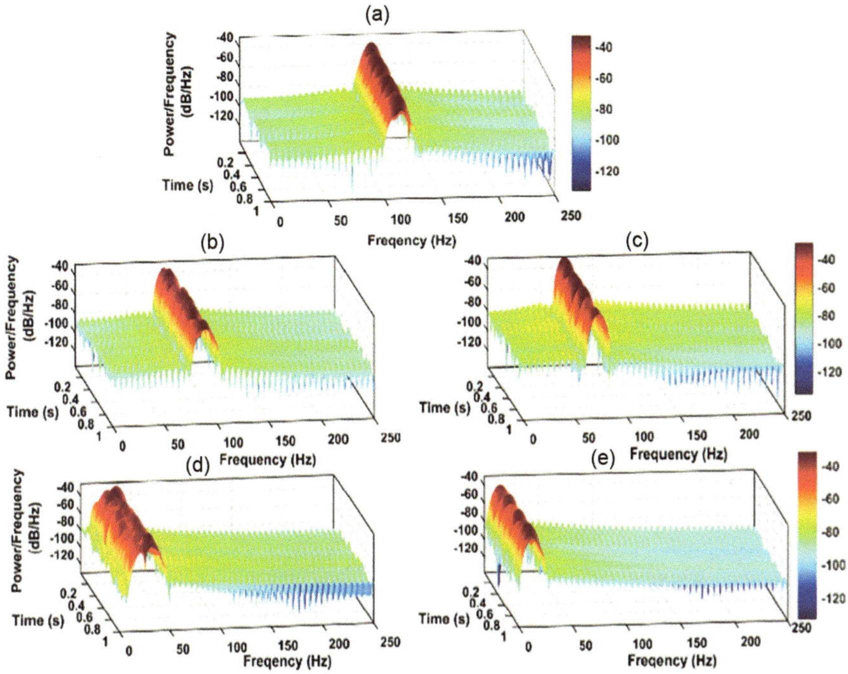

Figure 5.21 The 3D time-frequency spectrums of the mean of the signal components of sensor-10 under a white-noise for various states of the bridge including; (a) healthy, (b) 35% damaged, (c) 60% damaged, (d) 83% damaged, and (e) 100% damaged.

To assess the capability of the proposed method in addressing the location of damage, the acceleration signals recorded by the sensors 11, 9, 4, and 1 that represent the sensor locations with the nearest, the second near, far, and the farthest distances from to the damaged element, respectively, are analyzed. In this case, it is noteworthy that only a damage scenario with a 100% damaged element was considered to evaluate the performance of EWT-ANN in addressing the damage location which was conducted by removing a diagonal element from the truss as illustrated in Figure 5.14. By applying the EWT on the acceleration responses of the aforementioned sensors, the 12 components are obtained as depicted in Figure 5.22. The increments of the amplitude of the extracted component can be observed as the distance of the sensors from the damage location increases. Accordingly, from the observation on the spectrogram of the four selected sensors illustrated in Figure 5.23, the power increases in the lower frequency range as the distance of the analyzed sensor increases from the location of 100% damage.

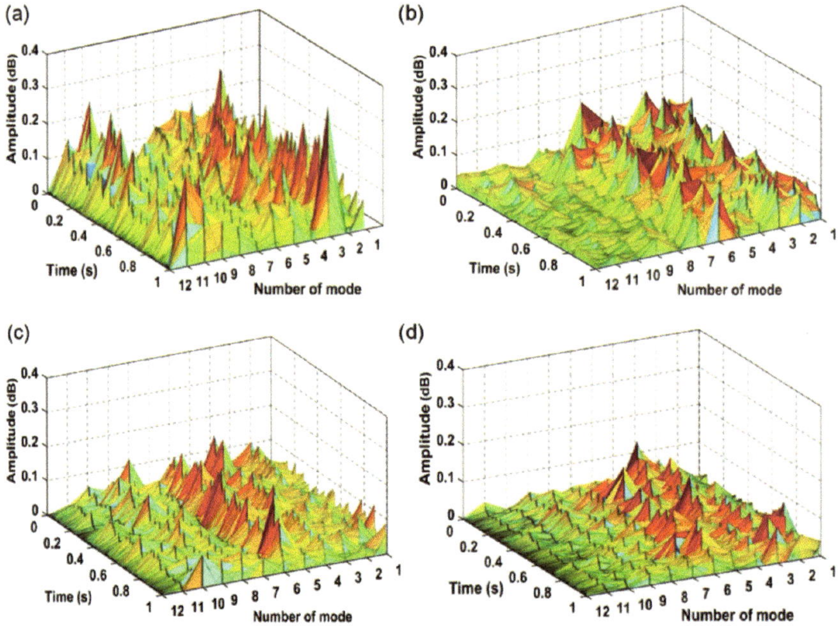

Figure 5.22 Signal mode components of different sensors placed at; (a) the nearest, (b) the second near, (c) far, and (d) the farthest distances from the damage location of the truss under a white-noise.

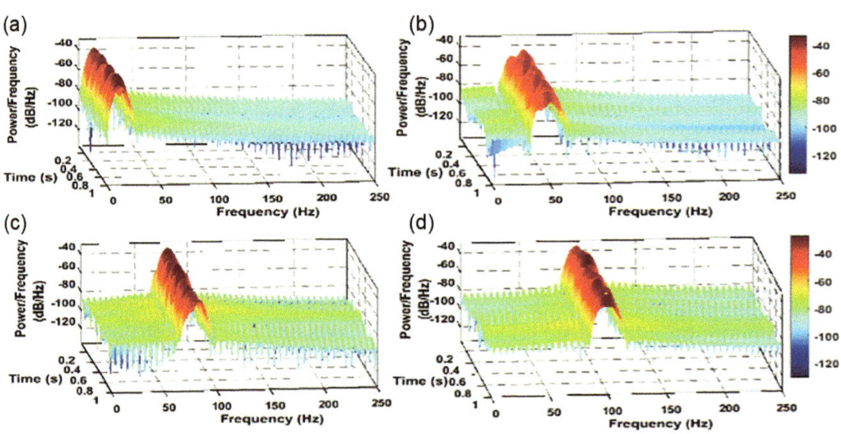

Figure 5.23 Time-frequency spectrograms of the signal component modes of different sensors placed at; (a) the nearest, (b) the second near, (c) far, and (d) the farthest distances from the damage location of the truss under white noise excitation.

5.4.4 Detecting the presence, severity, and location of damage under impact load excitation

To demonstrate the versatility of the procedure under discussion, the capability of the EWT-ANN approach in identifying the damage existence and location in the truss is also assessed when subjected to an impulsive impact load with a peak value of 100 N and a duration of 35 ms as introduced in Figure 5.14b. The impact load considered in this study can represent collisions of vessel deck-house with the lower chord of bridge superstructure during their deviations from the navigating routes in the real world. Similar to the reason mentioned in Section 4.1, the one-second acceleration response of an example checkpoint (here, sensor-10) for the healthy bridge is presented in Figure 5.24a. Furthermore, the corresponding mode components of the vibration response decomposed by the EWT are illustrated in Figure 5.24b. It should be noted that all the damage scenarios sketched

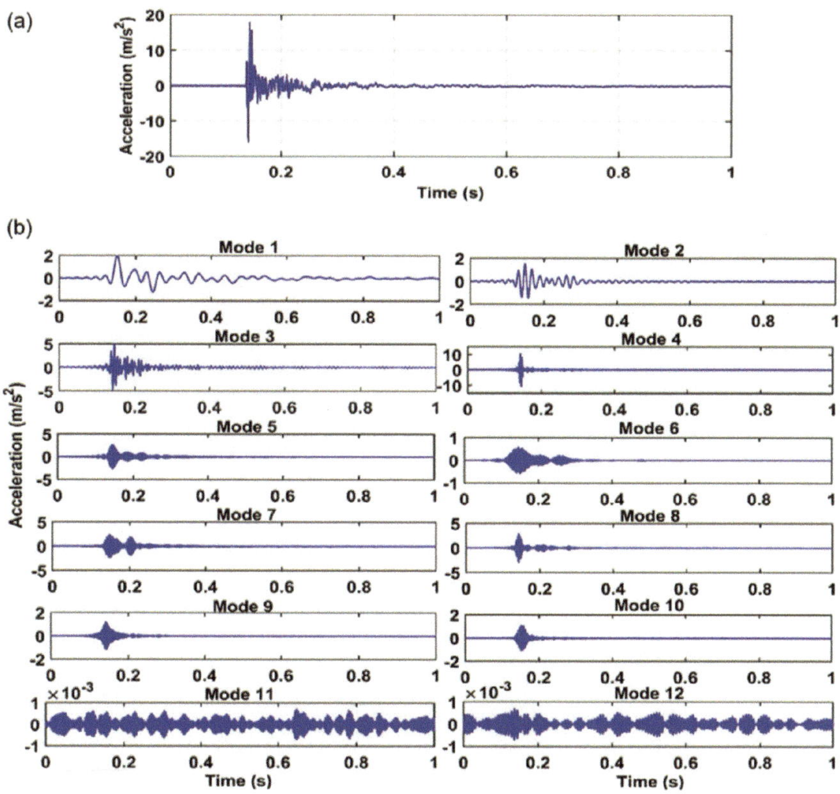

Figure 5.24 (a) Acceleration signal recorded by sensor-10 for the healthy bridge under an impact load, (b) the corresponding signal mode components extracted by the EWT.

in Section 4.1 are similarly considered in this section for the bridge under an impact load excitation.

Figure 5.25a illustrates the three-dimensional plots of the 12 mode components decomposed by the EWT for the healthy truss in comparison with those from the damaged states of the truss with severities of 35%, 60%, 83%, and 100% as shown in Figures 5.25b–e, respectively. It is observed that the amplitudes of signal modes especially between modes 2 and 4 significantly increase with the increase of the damage severity. Moreover, similar to the trend observed for damage scenarios of the bridge when exposed

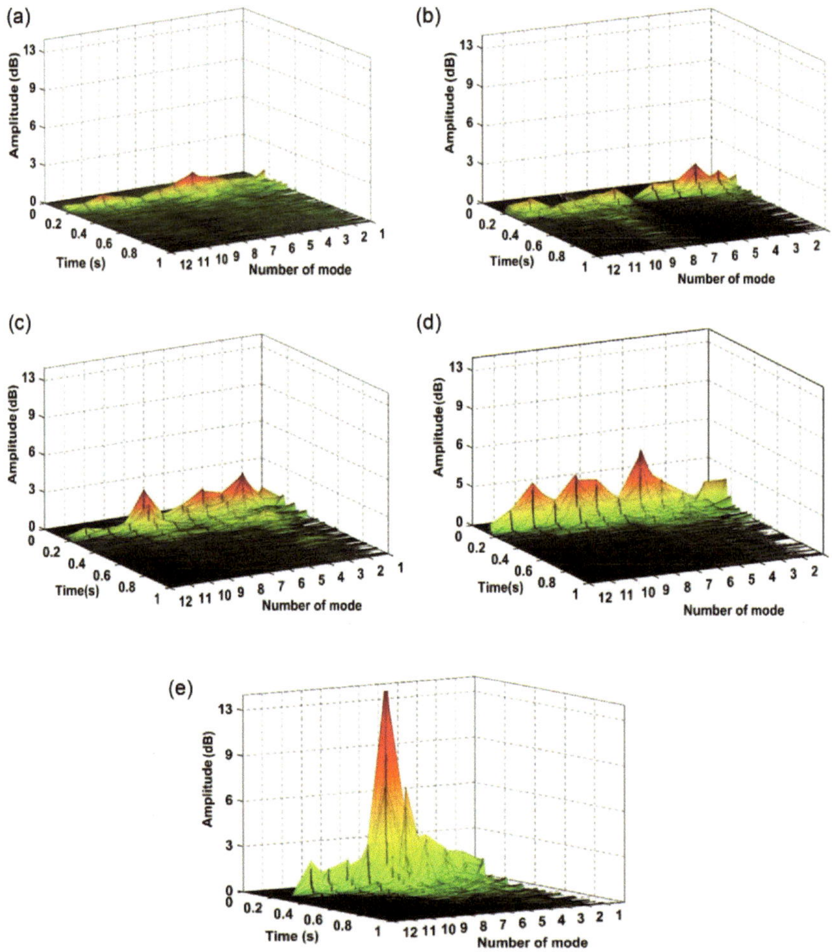

Figure 5.25 Signal mode components of sensor-10 by EWT for various states of the bridge under an impact load including; (a) healthy, (b) 35% damaged, (c) 60% damaged, (d) 83% damaged, and (e) 100% damaged.

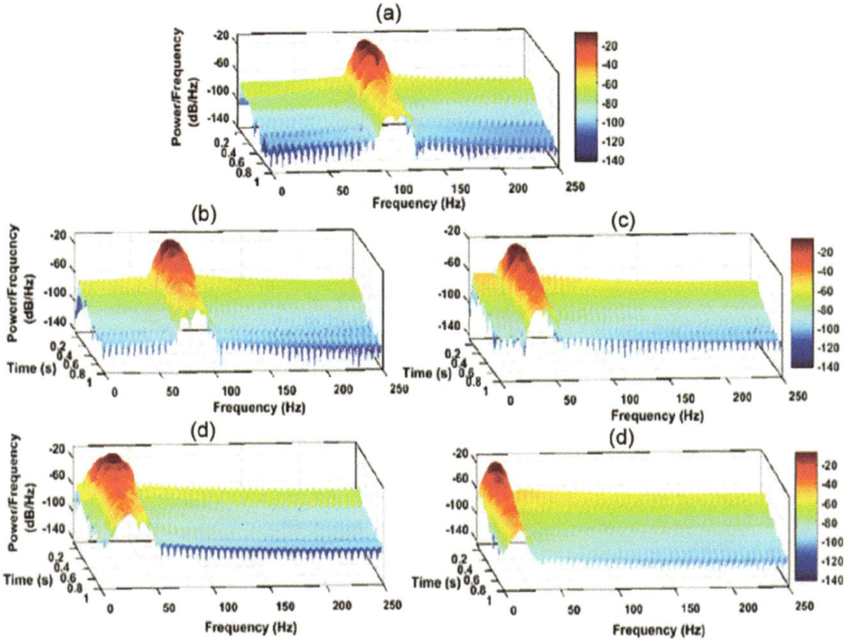

Figure 5.26 Time-frequency spectrograms of the signal mode components of sensor-10 for various states of the bridge under an impact load including; (a) healthy, (b) 35% damaged, (c) 60% damaged, (d) 83% damaged, and (e) 100% damaged.

to a white noise excitation, the frequency range of the enhanced power of the spectrums decreases with increasing the damage level as shown in Figure 5.26a–e. This is because the frequency range of the bridge response decreases in damaged states due to the stiffness reduction.

Similar to the dissections in Section 4.1, the vibration signals of sensors 11, 9, 4, and 1 that have the nearest, the second near, far, and the farthest distances relative to the damage location (with a 100% damage as identified in Figure 5.15), respectively, are analyzed under an impact excitation to address the damage location using the proposed approach. Figure 5.27a–d demonstrates the three-dimensional plots of the 12 mode components of these sensors. Similar to the observations for the bridge under a white noise excitation, the amplitude of the signal modes especially in modes between 2 and 4 significantly enhances with the decrease of the sensor distance from the damage location. Also, the power enhancement of the spectrums in the lower frequency range can be observed in Figure 5.28a–d by reducing the sensor location relative to the damaged element.

Figure 5.29a–e presents the results of the quantitative assessment of the proposed EWT-ANN approach in damage detection of the truss subjected to the impact load excitation based on the time-domain features including energy,

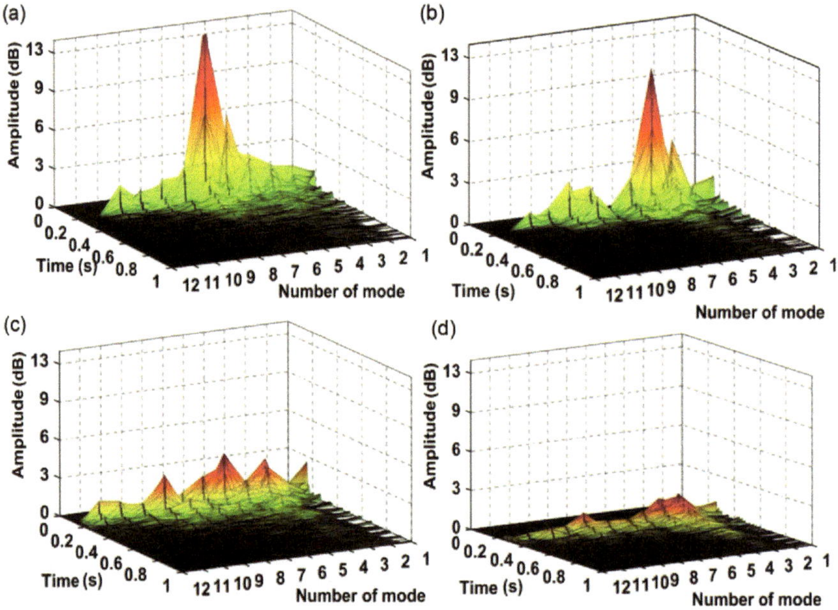

Figure 5.27 Mode components of the vibration signals of different sensors placed at; (a) the nearest, (b) the second near, (c) far, and (d) the farthest distances from the damage location of the truss under an impact load.

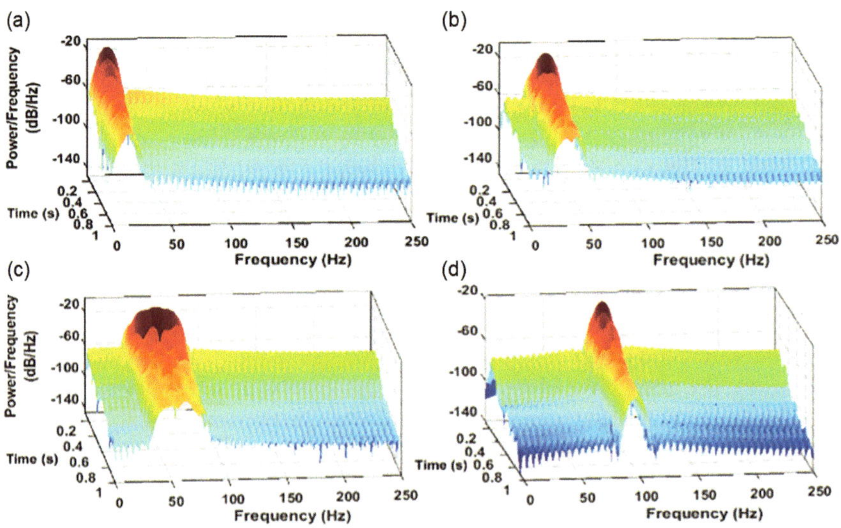

Figure 5.28 Time-frequency spectrograms of the component modes of vibration signals of different sensors placed at; (a) the nearest, (b) the second near, (c) the far, and (d) the farthest distances from the damaged element of the truss under an impact load.

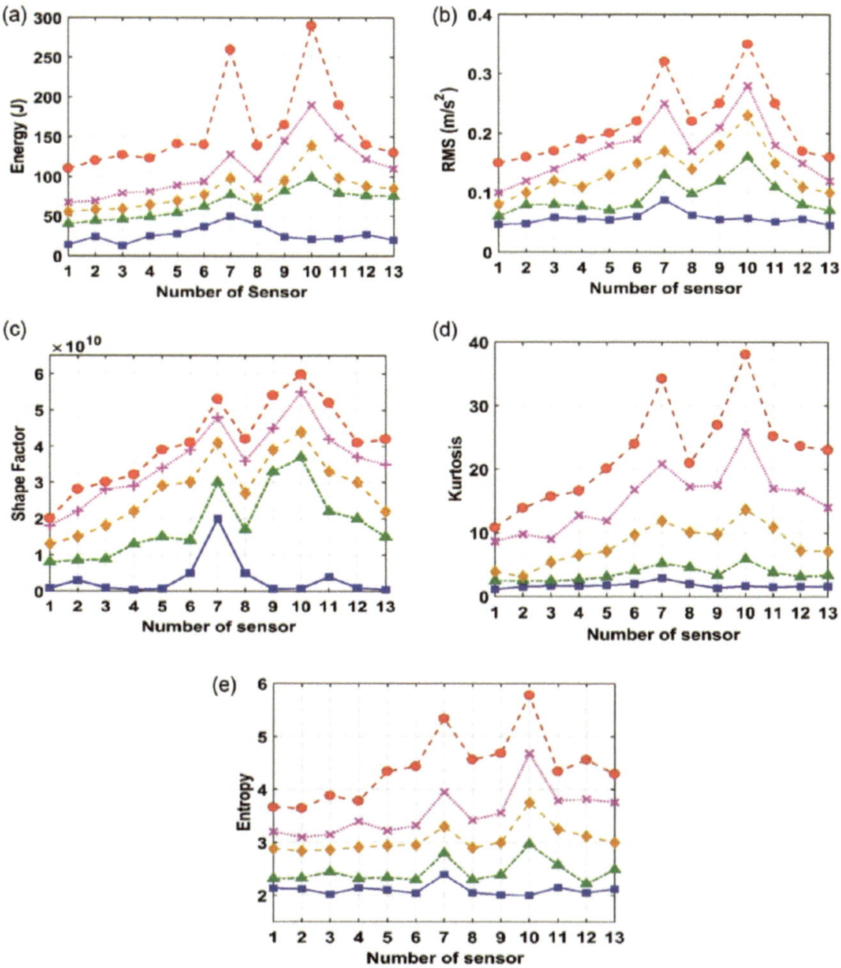

Figure 5.29 The estimated time-domain features by the EWT-ANN model of vibration signals from 13 sensors under an impact load to classify the damage levels including; (a) energy, (b) RMS, (c) shape factor, (d) kurtosis, and (e) entropy.

RMS, shape factor, kurtosis, and entropy. The trained EWT-ANN model of the healthy state of the truss bridge is separately examined by the extracted mode components in different damage levels of 35%, 60%, 83%, and 100% based on the aforementioned features. Except for the peak values of the indices that occurred for various states of the bridge at the location of the applied impact load around sensor-7, the features reach their peak values at the damage location (sensor-10). Moreover, the peak values of the features at the damage location significantly increase with enhancing the damage severity. Therefore, it can be seen that the proposed automated EWT-ANN approach based on the time-domain features can successfully and accurately detect and locate the damage in the bridge model exposed to different excitations.

5.5 CHAPTER SUMMARY

In the first part of this chapter, the performance of a combined CEEMDAN-Hilbert Transform-Artificial Neural Network (CEEMDAN-HT-ANN) model as a fusion of data analysis and machine learning approach was experimentally assessed in identifying the presence, location, and severity of the damage on a laboratory-model steel truss bridge. To this end, the CEEMDAN was used to decompose the response of the bridge subjected to a white noise excitation. Then, the HT technique was applied to the IMFs before and after the damage to extract the key signal parameters including the energy, IA, unwrapped phase, and IF. After employing HHT, the IMFs and extracted parameters were selected as the input and output layers of the ANN for the healthy state of the bridge, respectively. Then, the proposed CEEMDAN-HT-ANN model was tested by the IMFs of the damaged truss bridge. The estimated responses of the proposed model based on four parameters were qualitatively compared for different damage scenarios. Moreover, three damage indices were identified based on the estimated output of the proposed CEEMDAN-HT-ANN model under four damage severities and locations as the effective indicators to quantitatively classify the level and detect the location of the damage. By analyzing the experimental results, it was concluded that the damage features of the signals were accurately distinguished and extracted with the proposed approach. In addition, both quantitative and qualitative results showed the capability and robustness of the CEEMDAN-HT-ANN model in addressing the damage location, classifying the severity, and detecting the damage.

In the second part of this chapter, a combination of EWT and an artificial neural network named EWT-ANN is proposed based on several time-domain vibration signal features (energy, RMS, shape factor, kurtosis, and entropy) of a laboratory-scale model of a steel truss bridge exposed to a band-limited white noise or impact load excitations. The main advantage of the proposed approach is to provide an efficient automated damage-detection method to rapidly assess, identify, and classify the structural damage and nonlinear behavior of a laboratory-scale model of a steel truss bridge. Once the acceleration responses of the bridge are decomposed by the EWT into a series of the signal mode components as the first step in implementing the proposed method, these modes are used as the inputs in training an ANN. Additionally, the energy and four statistical time-domain features extracted from the components are employed in training the ANN as an output layer to detect the damage presence, intensity, and location. After training the model using the bridge response data in a healthy state exposed to white noise and impact load excitations, the EWT-ANN model acts as a structural database (i.e., target). Then, the extracted mode components from the acceleration signals of the bridge, in different damage states, are processed to assess the validity of the proposed approach. The quantitative differences between the output results of the EWT-ANN model prior to and post the damaged states of the bridge were utilized to quantify and localize the damage. From the evaluation of the five features for different damage states of the truss under white noise excitation, it was found that the magnitudes of all features increased about 100% at the location of damage (Sensor 10) compared to those at the farthest sensor location (Sensor 1). In addition, the sensitivities of the energy and RMS features to the damage severity were more pronounced when the damage severity was higher than 60%. As such, when the damage severity increased from 60% to 83%, the magnitudes of the energy and RMS features enhanced about 75% and 34%, respectively. However, the shape factor and kurtosis features presented lower sensitivity levels to the increase of damage severity. In other words, a damage severity of 35% significantly increased the magnitudes of shape factor (seven times) and kurtosis (two times) features around the damage location. Moreover, it was found the magnitudes of entropy feature significantly increased (35%) when the damage level reached 60%. Besides, when the truss was subjected to an impact loading excitation, although a damage level of 35% caused notable increases in the magnitudes of the energy, RMS, shape factor, and entropy features around the damage location, the kurtosis feature was not significantly sensitive to this range of damage. However, the magnitude of kurtosis feature increased about three times at the damage location when the damage severity reached 60%. Therefore, the experimental results quantitatively and qualitatively demonstrated the capability of the proposed approach in identifying the damage existence, intensity, and location.

Finally, Chapter 6 presents particular conclusions and main findings of this book and also recommendation for future works.

Chapter 6

Conclusions

6.1 SUMMARY OF THE PRESENT WORK

In this book, the performance of traditional and novel signal processing techniques for decomposing the acceleration response and dynamic feature extraction was experimentally assessed in the time-frequency domain to detect, classify, and locate structural damage of steel truss bridge laboratory model. Besides, one of the serious issues of traditional signal processing techniques in analyzing the responses of real-life structures is related to the presentation of fundamental information of nonlinear, non-stationary, and noisy signals with closely-spaced frequencies. To overcome this difficulty, numerous studies have been recently done to explore proper signal processing techniques to efficiently present high-resolution representations for nonlinear characteristics of analyzed signals. Despite existing extensive reviews on vibration-based signal processing techniques in time and frequency domains for SHM purposes, there exists no study in categorizing the signal processing techniques based on the feature extraction with time-frequency representations. To fill this gap, this book presents a comprehensive state-of-the-art review on the applications of signal processing techniques in time, frequency, and time-frequency domains that have been commonly used in structural health monitoring in various civil engineering structures when subjected to different excitations under environmental or controlled conditions for damage detection, localization, and quantification in various structural systems.

The progressive trend of time, frequency, and time-frequency analysis methods was reviewed by summarizing their advantages and disadvantages when utilized for different applications in various structural and mechanical systems as shown in Figures 6.1 and 6.2. From the review on time-domain techniques, it is found that these techniques are mostly used in linear systems while they are highly sensitive to noise. Among the time-domain techniques, ARMAX has more advantages in analyzing nonlinear and non-stationary signals compared to other models and it efficiently can mitigate the noise disturbance of vibration responses of structures. A summary of the

DOI: 10.1201/9781003499046-6

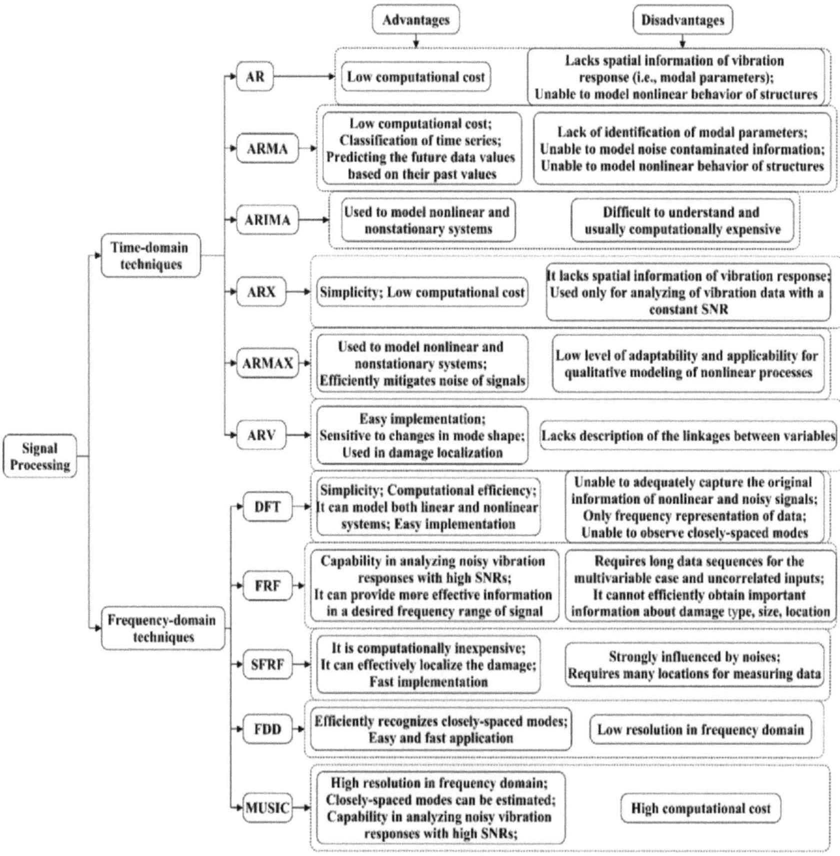

Figure 6.1 A brief summary of advantages and disadvantages associated with time-domain and frequency-domain signal processing techniques.

advantages and disadvantages of time-domain and frequency domain signal processing techniques is illustrated in Figure 6.1. Besides, from the review on frequency-based techniques, their substantial limitations were found in the analysis and presentation of data using frequency spectrums. After a careful review of these techniques, it was found that new techniques such as MUSIC should be assessed and compared with other frequency-based signal processing techniques such as traditional FFT, PSD, and FDD.

Moreover, the progressive trend of time-frequency analysis techniques in solving the issue associated with the performance of traditional signal processing techniques was reviewed by clarifying the advantages and disadvantages of each technique in Figure 6.2. It was found that despite many applications of EMD in the literature, this method suffers some drawbacks such as generating undesirable IMFs at the low-frequency

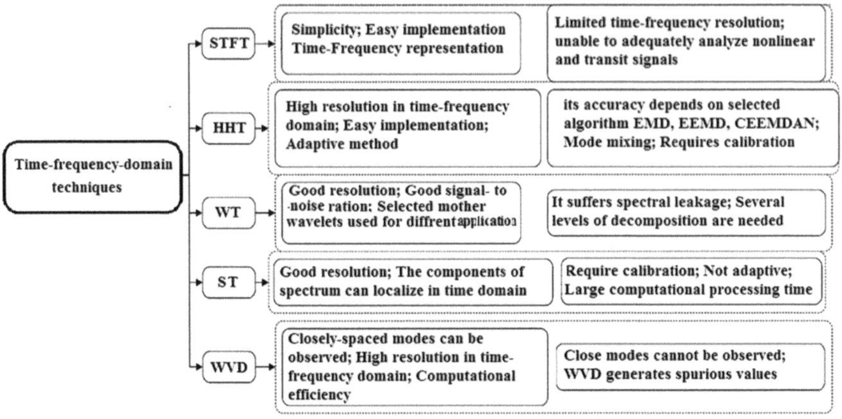

Figure 6.2 A brief summary of advantages and disadvantages associated with time-frequency domain signal processing techniques

region with the mode mixing problem leading to inaccurate SHM results. To overcome this, Wu and Huang [276] proposed the ensemble empirical mode decomposition which was an improved method by adding an identically white Gaussian noise with appropriate scale and standard deviation. EEMD can diminish the mode mixing problem, by adding Gaussian white noise to the input signal. However, The EEMD has two major drawbacks: (i) residual noise in IMF and (ii) difficulty in averaging of a different number of IMFs caused by adding different white Gaussian noise to the signal. To overcome this Torres et al. [278] solved these faults by proposing the CEEMDAN. The CEEMDAN has a different trend from EEMD in adding a particular noise at each step of the decomposition process instead of adding the white Gaussian noise which is obtained with a unique residue after each IMF's extraction.

Furthermore, it was found that the EWT technique which is known as an improved WT-based technique and needs to assess in different structures under various excitations. Also, the combination of EWT with ANN was not reported in SHM and damage detection in civil engineering. Besides, Since the use of the HHT method combined with neural networks is very limited in the literature, this book aims to evaluate the performance of a proposed methodology from the combination of CEEMDAN-Hilbert transform and neural network techniques called CEEMDAN-HT-ANN, which has not been reported for damage detection in civil engineering. Due to the potential advantages of the CEEMDAN demonstrated by Torres et al. [278] compared to the previous generation of EMD-based techniques, CEEMDAN is employed to decompose the acceleration response of a steel-truss bridge model.

After a comprehensive review in the literature, potential techniques that need more assessment in SHM and damage detection procedure were selected, and their combination with ANN of a laboratory-scaled steel truss bridge model was evaluated. The main findings of the proposed CEEMDAN-HT-ANN, CEEMDAN-MUSIC, and EWT-ANN models are as follows, respectively.

First, the performance of the EMD-based signal processing technique referred to as CEEMDAN was experimentally assessed in identifying the presence, location, and severity of the damage for a steel truss bridge model. Based on the evaluations using three signal processing techniques, including EMD, EEMD, and CEEMDAN, the first IMF extracted by the CEEMDAN was selected as the evaluation criterion in detecting the structural damage. In addition, four key parameters of the signal, including the energy, IA, unwrapped phase, and IF extracted through applying HT to the IMFs, are considered to investigate the existence, severity, and location of the damage in the model. Furthermore, the sensitivity of CEEMDAN compared to the previous generations of the EMD-based techniques is investigated by proposing several improved damage indices from the combinations of two statistical signal features, including kurtosis and entropy with the energy and IA features of the analyzed signal.

The main findings of this article from the analysis of the experimental acceleration response of the truss can be summarized as follows:

- The CEEMDAN approach is a novel extension of EMD which accurately reproduces a proceed signal and mitigates the mode mixing problem.
- The intensity of spikes of the first IMF decomposed by CEEMDAN increased more significantly compared to those from EEMD and EMD techniques, which demonstrate the completeness of CEEMDAN in decomposing the acceleration response of the structure. Therefore, the CEEMDAN presented a more sensitive damage detection approach compared to EMD and EEMD.
- Increasing the energy and IA values and decreasing the unwrapped phase values of the first IMF were observed with the appearance of the damage, with increasing the level of damage and with decreasing the distance of the desired sensor from the damage location.
- Although EMD and EEMD were able to detect the damage, a significant improvement of CEEMDAN compared to the other techniques in detecting the existence, severity, and location of the damage using the damage indices based on energy, IA, and unwrapped phase parameters was concluded.
- By assessing the IF of IMFs in the time–frequency–energy domain, an increase in the power of the frequency in the low ranges is observed when the damage appears. Furthermore, this technique can provide an indication of the general damage region and being a sensitive indicator of damage.

- By comparing the value of the damage indices based on energy, IA, and unwrapped phase features, it was found that the energy feature represents a better approach in detecting, locating, and classifying the severity of the damage.
- It was concluded that the proposed damage indices based on the combinations of the kurtosis and entropy features with the energy and IA features resulted in more sensitive indices in identifying the presence, intensity, and location of the damage compared to those based on the direct utilizing of the energy and IA features. In addition, more sensitive indices were obtained when combining the entropy feature with the energy and IA compared to those in which the kurtosis feature was utilized.

Furthermore, the use of machine learning technique speeds the process up, and fill the gap in knowledge of ANN-aided novel signal processing in damage detection procedure. To this end, the performance of a combined CEEMDAN-Hilbert Transform-Artificial Neural Network (CEEMDAN-HT-ANN) model as a fusion of data analysis and machine learning approach was experimentally assessed in identifying the presence, location, and severity of the damage on a laboratory-model steel truss bridge. To this end, the CEEMDAN was used to decompose the response of the bridge subjected to a white noise excitation. Then, the HT technique was applied to the IMFs before and after the damage to extract the key signal parameters including the energy, instantaneous amplitude (IA), unwrapped phase, and instantaneous frequency (IF). After employing HHT, the IMFs and extracted parameters were selected as the input and output layers of the ANN for the healthy state of the bridge, respectively. Then, the proposed CEEMDAN-HT-ANN model was tested by the IMFs of the damaged truss bridge. The estimated responses of the proposed model based on four parameters were qualitatively compared for different damage scenarios. Moreover, three damage indices were identified based on the estimated output of the proposed CEEMDANHT-ANN model under four damage severities and locations as the effective indicators to quantitatively classify the level and detect the location of the damage. By analyzing the experimental results, it was concluded that the damage features of the signals were accurately distinguished and extracted with the proposed approach. In addition, both quantitative and qualitative results showed the capability and robustness of the CEEMDAN-HT-ANN model in addressing the damage location, classifying the severity, and detecting the damage.

Moreover, compared to the traditional frequency-based signal processing techniques such as FFT, PSD, and FDD, the MUSIC technique has advantages in signal processing including the presentation of results in the frequency domain with a high-resolution, estimation of closely-spaced modes, and employing high SNR. Therefore, the use of the MUSIC-based techniques is highly recommended for future research works. To this end, the performance of a combined CEEMDAN-MUSIC technique as a novel signal processing technique in identifying the presence, addressing the location, and classifying the level of damage in a laboratory-scale model of a steel truss bridge was experimentally assessed. The implementation process of this technique begins by employing the CEEMDAN algorithm to decompose the acceleration response of the truss bridge exposed to a white noise excitation. Then, the MUSIC technique was applied to the first IMF before and after damage to compute the power density pseudospectrum. In addition, the absolute differences of the spectral peaks in the frequency (Df) and power magnitude (Dp) ranges extracted from the responses of different damage scenarios varying in terms of the damage severity and the damage location are considered as the effective indicators to classify the level and detect the location of the damage. Significant improvements were observed in the damage detection and characterization process of the truss using the proposed method by comparing the results from the indicator values of the combined CEEMDAN-MUSIC algorithm with those from pure MUSIC technique and several frequency domain techniques such as FFT, PSD, and FDD. As such, compared to many spurious peaks captured by the conventional frequency domain techniques leading to notable difficulties in the damage detection process, the proposed method was clearly able to address the classification of the damage severity levels and the damage location, as well as the perception of damages at lower levels. As a result, the method can be recommended for health monitoring of various structures according to the proposed framework in this study.

Finally, the combination of empirical wavelet transform and artificial neural network named EWT-ANN is proposed based on several time-domain vibration signal features (energy, RMS, shape factor, kurtosis, and entropy) of a laboratory-scale model of a steel truss bridge exposed to a band-limited white noise or impact load excitations. The main advantage of the proposed approach is to provide an efficient automated damage-detection method to rapidly assess, identify, and classify the structural damage and nonlinear behavior of a laboratory-scale model of a steel truss bridge. Once the acceleration responses of the bridge are decomposed by the EWT into a series of signal mode components as the first step in implementing the proposed method, these modes are used as the inputs in training an ANN. Additionally, the energy and four statistical time-domain features extracted from the components are employed in training the ANN as an output layer to detect the damage presence, intensity, and location. After training the model using the bridge response data in a healthy state exposed to white noise and impact load excitations, the EWT-ANN model acts as a structural database (i.e., target). Then, the extracted mode components from the acceleration signals of the bridge, in different damage states, are processed to assess the validity of the proposed approach. The quantitative differences between the output results of the EWT-ANN model prior to and post the damaged states of the bridge were utilized to quantify and localize the damage. From the evaluation of the five features for different damage states of the truss under white noise excitation, it was found that the magnitudes of all features increased about 100% at the location of damage (Sensor 10) compared to those at the farthest sensor location (Sensor 1). In addition, the sensitivities of the energy and RMS features to the damage severity were more pronounced when the damage severity was higher than 60%. As such, when the damage severity increased from 60% to 83%, the magnitudes of the energy and RMS features enhanced by about 75% and 34%, respectively. However, the shape factor and kurtosis features presented lower sensitivity levels to the increase of damage severity. In other words, a damage severity of 35% significantly increased the magnitudes of shape factor (seven times) and kurtosis (two times) features around the damage location. Moreover, it was found the magnitudes of entropy feature significantly increased (35%) when the damage level reached 60%. Besides, when the truss was subjected to an impact loading excitation, although a damage level of 35% caused notable increases in the magnitudes of the energy, RMS, shape factor, and entropy features around the damage location, the kurtosis feature was not significantly sensitive to this range of damage. However, the magnitude of kurtosis feature increased about three times at the damage location when the damage severity reached 60%. Therefore, the experimental results quantitatively and qualitatively demonstrated the capability of the proposed approach in identifying the damage existence, intensity, and location.

6.2 RECOMMENDATIONS FOR FUTURE WORKS

Compared to the traditional frequency-based signal processing techniques, the MUSIC technique has advantages in signal processing including the presentation of results in the frequency domain with a high-resolution, estimation of closely-spaced modes, and employing high SNR. Therefore, the use of the MUSIC-based techniques is highly recommended for future research works. Moreover, EWT has a serious limitation in processing noisy signals which contain overlapping segments of the Fourier spectrum. To overcome this deficiency, it was recommended to hybridize the use of EWT and improve frequency-based techniques such as MUSIC algorithm which is efficiently able to identify the signal components with certain boundaries.

References

1. Amezquita-Sanchez, J. P., & Adeli, H. (2016). Signal processing techniques for vibration-based health monitoring of smart structures. *Archives of Computational Methods in Engineering*, *23*(1), 1–15.
2. Caesarendra, W., & Tjahjowidodo, T. (2017). A review of feature extraction methods in vibration-based condition monitoring and its application for degradation trend estimation of low-speed slew bearing. *Machines*, *5*(4), 21.
3. Goyal, D., & Pabla, B. S. (2016). The vibration monitoring methods and signal processing techniques for structural health monitoring: A review. *Archives of Computational Methods in Engineering*, *23*(4), 585–594.
4. Sarmadi, H., Entezami, A., Saeedi Razavi, B., & Yuen, K. V. (2021). Ensemble learning-based structural health monitoring by Mahalanobis distance metrics. *Structural Control and Health Monitoring*, *28*(2), e2663.
5. Entezami, A., Shariatmadar, H., & Mariani, S. (2020). Early damage assessment in large-scale structures by innovative statistical pattern recognition methods based on time series modeling and novelty detection. *Advances in Engineering Software*, *150*, 102923.
6. Entezami, A., Sarmadi, H., Salar, M., De Michele, C., & Arslan, A. N. (2020). A novel data-driven method for structural health monitoring under ambient vibration and high-dimensional features by robust multidimensional scaling. *Structural Health Monitoring*. doi: 10.1177/1475921720973953.
7. Bursi, O. S., Kumar, A., Abbiati, G., & Ceravolo, R. (2014). Identification, model updating, and validation of a steel twin deck curved cable-stayed footbridge. *Computer-Aided Civil and Infrastructure Engineering*, *29*(9), 703–722.
8. Ghosh-Dastidar, S., & Adeli, H. (2003). Wavelet-clustering-neural network model for freeway incident detection. *Computer-Aided Civil and Infrastructure Engineering*, *18*(5), 325–338.
9. Jiang, X., & Adeli, H. (2004). Object-oriented model for freeway work zone capacity and queue delay estimation. *Computer-Aided Civil and Infrastructure Engineering*, *19*(2), 144–156.
10. Jiang, X., & Adeli, H. (2004). Wavelet packet-autocorrelation function method for traffic flow pattern analysis. *Computer-Aided Civil and Infrastructure Engineering*, *19*(5), 324–337.

11. Castillo, E., Calvino, A., Nogal, M., & Lo, H. K. (2014). On the probabilistic and physical consistency of traffic random variables and models. *Computer-Aided Civil and Infrastructure Engineering*, *29*(7), 496–517.
12. Haijema, R., & Hendrix, E. M. (2014). Traffic responsive control of intersections with predicted arrival times: A Markovian approach. *Computer-Aided Civil and Infrastructure Engineering*, *29*(2), 123–139.
13. Kannappan, L. (2009). Damage detection in structures using natural frequency measurements. University of New South Wales, Australian Defence Force Academy, School of Aerospace, Civil and Mechanical Engineering.
14. Kaveh, A., & Zolghadr, A. (2015). An improved CSS for damage detection of truss structures using changes in natural frequencies and mode shapes. *Advances in Engineering Software*, *80*, 93–100.
15. Natke, H. G., & Cempel, C. (2012). *Model-Aided Diagnosis of Mechanical Systems: Fundamentals, Detection, Localization, Assessment.* Heidelberg: Springer-Verlag.
16. Basseville, M., Abdelghani, M., & Benveniste, A. (2000). Subspace-based fault detection algorithms for vibration monitoring. *Automatica*, *36*(1), 101–109.
17. Sohn, H., & Farrar, C. R. (2000, June). Statistical process control and projection techniques for structural health monitoring. In *Proceedings of the European COST F3 Conference on System Identification and Structural Health Monitoring*, Madrid, Spain.
18. Fugate, M. L., Sohn, H., & Farrar, C. R. (2001). Vibration-based damage detection using statistical process control. *Mechanical Systems and Signal Processing*, *15*(4), 707–721.
19. Yan, A. M., De Boe, P., & Golinval, J. C. (2004). Structural damage diagnosis by Kalman model based on stochastic subspace identification. *Structural Health Monitoring*, *3*(2), 103–119.
20. Das, S., Saha, P., & Patro, S. K. (2016). Vibration-based damage detection techniques used for health monitoring of structures: A review. *Journal of Civil Structural Health Monitoring*, *6*(3), 477–507.
21. Sohn, H., Czarnecki, J. A., & Farrar, C. R. (2000). Structural health monitoring using statistical process control. *Journal of Structural Engineering*, *126*(11), 1356–1363.
22. Figueiredo, E., Figueiras, J., Park, G., Farrar, C. R., & Worden, K. (2011). Influence of the autoregressive model order on damage detection. *Computer-Aided Civil and Infrastructure Engineering*, *26*(3), 225–238.
23. Vamvoudakis-Stefanou, K. J., Sakellariou, J. S., & Fassois, S. D. (2018). Vibration-based damage detection for a population of nominally identical structures: unsupervised Multiple Model (MM) statistical time series type methods. *Mechanical Systems and Signal Processing*, *111*, 149–171.
24. Jayawardhana, M., Zhu, X., Liyanapathirana, R., & Gunawardana, U. (2015). Statistical damage sensitive feature for structural damage detection using AR model coefficients. *Advances in Structural Engineering*, *18*(10), 1551–1562.
25. De Lautour, O. R., & Omenzetter, P. (2010). Damage classification and estimation in experimental structures using time series analysis and pattern recognition. *Mechanical Systems and Signal Processing*, *24*(5), 1556–1569.

26. Gul, M., & Catbas, F. N. (2009). Statistical pattern recognition for Structural Health Monitoring using time series modeling: Theory and experimental verifications. *Mechanical Systems and Signal Processing*, *23*(7), 2192–2204.
27. Gharehbaghi, V. R., Nguyen, A., Farsangi, E. N., & Yang, T. Y. (2020). Supervised damage and deterioration detection in building structures using an enhanced autoregressive time-series approach. *Journal of Building Engineering*, *30*, 101292.
28. Pamwani, L., Agarwal, V., & Shelke, A. (2019). Damage classification and feature extraction in steel moment-resisting frame using time-varying autoregressive model. *Journal of Nondestructive Evaluation, Diagnostics and Prognostics of Engineering Systems*, *2*(2), 021002.
29. Cheng, J. J., Guo, H. Y., & Wang, Y. S. (2017). Structural nonlinear damage detection method using AR/ARCH model. *International Journal of Structural Stability and Dynamics*, *17*(08), 1750083.
30. Mei, L., Li, H., Zhou, Y., Li, D., Long, W., & Xing, F. (2020). Output-only damage detection of shear building structures using an autoregressive model-enhanced optimal subpattern assignment metric. *Sensors*, *20*(7), 2050.
31. Carden, E. P., & Brownjohn, J. M. (2008). ARMA modelled time-series classification for structural health monitoring of civil infrastructure. *Mechanical Systems and Signal Processing*, *22*(2), 295–314.
32. Nair, K. K., Kiremidjian, A. S., & Law, K. H. (2006). Time series-based damage detection and localization algorithm with application to the ASCE benchmark structure. *Journal of Sound and Vibration*, *291*(1-2), 349–368.
33. Zheng, H., & Mita, A. (2009). Localized damage detection of structures subject to multiple ambient excitations using two distance measures for autoregressive models. *Structural Health Monitoring*, *8*(3), 207–222.
34. Zheng, H., & Mita, A. (2008). Damage indicator defined as the distance between ARMA models for structural health monitoring. *Structural Control and Health Monitoring: The Official Journal of the International Association for Structural Control and Monitoring and of the European Association for the Control of Structures*, *15*(7), 992–1005.
35. Krishnan Nair, K., & Kiremidjian, A. S. (2007). Time series based structural damage detection algorithm using Gaussian mixtures modeling. doi:10.1115/1.2718241.
36. Shi, H., Worden, K., & Cross, E. J. (2019). A cointegration approach for heteroscedastic data based on a time series decomposition: An application to structural health monitoring. *Mechanical Systems and Signal Processing*, *120*, 16–31.
37. Zhang, X., Li, D., & Song, G. (2018). Structure damage identification based on regularized ARMA time series model under environmental excitation. *Vibration*, *1*(1), 138–156.
38. Bao, C., Hao, H., & Li, Z. X. (2013). Integrated ARMA model method for damage detection of subsea pipeline system. *Engineering Structures*, *48*, 176–192.
39. Georgantopoulou, A. S., & Fassois, S. D. (2014, November). Stationary or non-stationary random excitation for vibration-based structural damage detection? An exploratory study. In *Proceedings of the 6th International Symposium on NDT in Aerospace*, Madrid, Spain (pp. 12–14).

40. Omenzetter, P., & Brownjohn, J. M. W. (2006). Application of time series analysis for bridge monitoring. *Smart Materials and Structures, 15*(1), 129.
41. Xin, J., Zhou, J., Yang, S. X., Li, X., & Wang, Y. (2018). Bridge structure deformation prediction based on GNSS data using Kalman-ARIMA-GARCH model. *Sensors, 18*(1), 298.
42. Chen, Y., Corr, D. J., & Durango-Cohen, P. L. (2014). Analysis of common-cause and special-cause variation in the deterioration of transportation infrastructure: A field application of statistical process control for structural health monitoring. *Transportation Research Part B: Methodological, 59*, 96–116.
43. Posenato, D., Kripakaran, P., Inaudi, D., & Smith, I. F. (2010). Methodologies for model-free data interpretation of civil engineering structures. *Computers & Structures, 88*(7–8), 467–482.
44. Ling, Y., & Mahadevan, S. (2012). Integration of structural health monitoring and fatigue damage prognosis. *Mechanical Systems and Signal Processing, 28*, 89–104.
45. Kosorus, H., Honigl, J., & Kung, J. (2011, Toulouse, France August). Using R, WEKA and RapidMiner in time series analysis of sensor data for structural health monitoring. In *2011 22nd International Workshop on Database and Expert Systems Applications* (pp. 306–310). IEEE.
46. Bernal, D., Zonta, D., & Pozzi, M. (2012). ARX residuals in damage detection. *Structural Control and Health Monitoring, 19*(4), 535–547.
47. Yang, M., & Makis, V. (2010). ARX model-based gearbox fault detection and localization under varying load conditions. *Journal of Sound and Vibration, 329*(24), 5209–5221.
48. Oh, C. K., & Sohn, H. (2009). Damage diagnosis under environmental and operational variations using unsupervised support vector machine. *Journal of Sound and Vibration, 325*(1–2), 224–239.
49. Gul, M., & Catbas, F. N. (2011). Structural health monitoring and damage assessment using a novel time series analysis methodology with sensor clustering. *Journal of Sound and Vibration, 330*(6), 1196–1210.
50. Sohn, H., Farrar, C. R., Hunter, N. F., & Worden, K. (2001). Structural health monitoring using statistical pattern recognition techniques. *Journal of Dynamic Systems, Measurement, and Control, 123*(4), 706–711.
51. Li, H., & Ren, Y. (2018). Design and implementation of building structure monitoring system based on radio frequency identification (RFID). *International Journal of RF Technologies, 9*(1–2), 37–49.
52. Fasel, T. R., Sohn, H., Park, G., & Farrar, C. R. (2003, January). Application of frequency domain arx models and extreme value statistics to impedance-based damage detection. In *ASME International Mechanical Engineering Congress and Exposition*, Washington, DC (Vol. 37076, pp. 289–297).
53. Lakshmi, K., & Rama Mohan Rao, A. (2014). A robust damage-detection technique with environmental variability combining time-series models with principal components. *Nondestructive Testing and Evaluation, 29*(4), 357–376.
54. Krishnasamy, L., Arumulla, R. M. R., & Naga, G. (2018). An improved damage diagnostic technique based on Singular Spectrum Analysis and time series models. *Structure and Infrastructure Engineering, 14*(10), 1412–1431.

55. Poulimenos, A. G., & Sakellariou, J. S. (2019). A transmittance-based methodology for damage detection under uncertainty: An application to a set of composite beams with manufacturing variability subject to impact damage and varying operating conditions. *Structural Health Monitoring*, *18*(1), 318–333.
56. Ugalde, U., Anduaga, J., Martinez, F., & Iturrospe, A. (2015). SHM method for damage localization based on substructuring and VARX models. arXiv preprint arXiv:1501.01905.
57. Silva, S. D., Dias Júnior, M., & Lopes Junior, V. (2007). Damage detection in a benchmark structure using AR-ARX models and statistical pattern recognition. *Journal of the Brazilian Society of Mechanical Sciences and Engineering*, *29*(2), 174–184.
58. Tatsis, K., Dertimanis, V., Ou, Y., & Chatzi, E. (2020). GP-ARX-Based structural damage detection and localization under varying environmental conditions. *Journal of Sensor and Actuator Networks*, *9*(3), 41.
59. Wei, Z., Yam, L. H., & Cheng, L. (2005). NARMAX model representation and its application to damage detection for multi-layer composites. *Composite Structures*, *68*(1), 109–117.
60. Lakshmi, K., Rao, A. R. M., & Gopalakrishnan, N. (2017). Singular spectrum analysis combined with ARMAX model for structural damage detection. *Structural Control and Health Monitoring*, *24*(9), e1960.
61. Do, N. T., & Gül, M. (2020). Structural damage detection under multiple stiffness and mass changes using time series models and adaptive zero-phase component analysis. *Structural Control and Health Monitoring*, *27*(8), e2577.
62. Lakshmi, K., & Rao, A. R. M. (2016). Structural damage detection using ARMAX time series models and cepstral distances. *Sādhanā*, *41*(9), 1081–1097.
63. Mei, L., Li, H., Zhou, Y., Wang, W., & Xing, F. (2019). Substructural damage detection in shear structures via ARMAX model and optimal subpattern assignment distance. *Engineering Structures*, *191*, 625–639.
64. Xie, L., & Mita, A. (2017). An innovative substructure damage identification approach for shear structures based on ARMAX models. *Procedia Engineering*, *188*, 119–124.
65. Gislason, G. P., Mei, Q., & Gül, M. (2019). Rapid and automated damage detection in buildings through ARMAX analysis of wind induced vibrations. *Frontiers in Built Environment*, *5*, 16.
66. Saaed, T. E., & Nikolakopoulos, G. (2016). Identification of building damage using ARMAX model: A parametric study. *Diagnostyka*, *17*, 3–14.
67. Xing, Z., & Mita, A. (2012). A substructure approach to local damage detection of shear structure. *Structural Control and Health Monitoring*, *19*(2), 309–318.
68. Li, C. S., Ko, W. J., Lin, H. T., & Shyu, R. J. (1993). Vector autoregressive modal analysis with application to ship structures. *Journal of Sound and Vibration*, *167*(1), 1–15.
69. Mosavi, A. A., Dickey, D., Seracino, R., & Rizkalla, S. (2012). Identifying damage locations under ambient vibrations utilizing vector autoregressive models and Mahalanobis distances. *Mechanical Systems and Signal Processing*, *26*, 254–267.

70. Lakshmi, K., & Rao, A. R. M. (2014). A robust SHM scheme combining time series models with dynamic QPSO algorithm. *Procedia Engineering, 86*, 870–877.
71. Mattson, S. G., & Pandit, S. M. (2006). Statistical moments of autoregressive model residuals for damage localisation. *Mechanical Systems and Signal Processing, 20*(3), 627–645.
72 Mattson, S. G., & Pandit, S. M. (2006). Damage detection and localization based on outlying residuals. *Smart Materials and Structures, 15*(6), 1801.
73. Okasha, N. M., Frangopol, D. M., Saydam, D., & Salvino, L. W. (2011). Reliability analysis and damage detection in high-speed naval craft based on structural health monitoring data. *Structural Health Monitoring, 10*(4), 361–379.
74. Bodeux, J. B., & Golinval, J. C. (2001). Application of ARMAV models to the identification and damage detection of mechanical and civil engineering structures. *Smart Materials and Structures, 10*(3), 479.
75. Mosavi, A. A., Dickey, D., Seracino, R., & Rizkalla, S. H. (2010, April). Time-series models for identifying damage location in structural members subjected to ambient vibrations. In *Health Monitoring of Structural and Biological Systems 2010* (Vol. 7650, p. 76502N). International Society for Optics and Photonics. doi: 10.1117/12.849048.
76. Lakshmi, K., & Rao, A. (2015). Damage identification technique based on time series models for LANL and ASCE benchmark structures. *Insight-Non-Destructive Testing and Condition Monitoring, 57*(10), 580–588.
77. Kraemer, P., & Fritzen, C. P. (2011). Aspects of operational modal analysis for structures of offshore wind energy plants. In T. Proulx, (ed.) *Structural Dynamics and Renewable Energy*, Volume 1 (pp. 145–152). New York: Springer.
78. Sinou, J. J. (2009). A review of damage detection and health monitoring of mechanical systems from changes in the measurement of linear and nonlinear vibrations.
79. Adams, R. D., Cawley, P., Pye, C. J., Stone, B. J., & Davies, W. G. R. (1979). A vibration technique for nondestructively assessing the integrity of structures. *Journal of Mechanical Engineering Science, 21*(1), 57.
80. Stubbs, N., & Osegueda, R. (1990). Global damage detection in solids- Experimental verification. *International Journal of Analytical and Experimental Modal Analysis, 5*, 81–97.
81. Doebling, S. W., Farrar, C. R., & Prime, M. B. (1998). A summary review of vibration-based damage identification methods. *Shock and Vibration Digest, 30*(2), 91–105.
82. Nikolakopoulos, P. G., Katsareas, D. E., & Papadopoulos, C. A. (1997). Crack identification in frame structures. *Computers & Structures, 64*(1–4), 389–406.
83. Salawu, O. S. (1997). Detection of structural damage through changes in frequency: A 7. review. *Engineering Structures, 19*(9), 718–723.
84. Williams, E. J., & Messina, A. (1999). Applications of the multiple damage location assurance criterion. In *Key Engineering Materials* (Vol. 167, pp. 256–264). Trans Tech Publications Ltd. 10.4028/www.scientific.net/KEM.167-168.256.
85. Morassi, A. (2001). Identification of a crack in a rod based on changes in a pair of natural frequencies. *Journal of Sound and Vibration, 242*(4), 577–596.

86. Nagayama, T., Reksowardojo, A. P., Su, D., & Mizutani, T. (2017). Bridge natural frequency estimation by extracting the common vibration component from the responses of two vehicles. *Engineering Structures*, *150*, 821–829.
87. Soh, C. K., Tseng, K. K., Bhalla, S., & Gupta, A. (2000). Performance of smart piezoceramic patches in health monitoring of a RC bridge. *Smart materials and Structures*, *9*(4), 533.
88. Aktan, A. E., Catbas, F. N., Ciloglu, S. K., & Grimmelsman, Q. P. (2005). Opportunities and challenges in health monitoring of constructed systems by modal analysis.
89. Lynch, J. P., Wang, Y., Loh, K. J., Yi, J. H., & Yun, C. B. (2006). Performance monitoring of the Geumdang Bridge using a dense network of high-resolution wireless sensors. *Smart Materials and Structures*, *15*(6), 1561.
90. Comanducci, G., Ubertini, F., & Materazzi, A. L. (2015). Structural health monitoring of suspension bridges with features affected by changing wind speed. *Journal of Wind Engineering and Industrial Aerodynamics*, *141*, 12–26.
91. Yarnold, M. T., & Moon, F. L. (2015). Temperature-based structural health monitoring baseline for long-span bridges. *Engineering Structures*, *86*, 157–167.
92. Zong, Z., Lin, X., & Niu, J. (2015). Finite element model validation of bridge based on structural health monitoring-Part I: Response surface-based finite element model updating. *Journal of Traffic and Transportation Engineering (English Edition)*, *2*(4), 258–278.
93. Rolek, P., Bruni, S., & Carboni, M. (2016). Condition monitoring of railway axles based on low frequency vibrations. *International Journal of Fatigue*, *86*, 88–97.
94. Brincker, R., Zhang, L., & Andersen, P. (2001). Modal identification of output-only systems using frequency domain decomposition. *Smart Materials and Structures*, *10*(3), 441.
95. Yuen, K. V., & Katafygiotis, L. S. (2005). Model updating using noisy response measurements without knowledge of the input spectrum. *Earthquake Engineering & Structural Dynamics*, *34*(2), 167–187.
96. Cheraghi, N., Zou, G. P., & Taheri, F. (2005). Piezoelectric-based degradation assessment of a pipe using fourier and wavelet analyses. *Computer-Aided Civil and Infrastructure Engineering*, *20*(5), 369–382.
97. Lee, J., & Kim, S. (2007). Structural damage detection in the frequency domain using neural networks. *Journal of Intelligent Material Systems and Structures*, *18*(8), 785–792.
98. Amezquita-Sanchez, J. P., Osornio-Rios, R. A., Romero-Troncoso, R. J., & Dominguez-Gonzalez, A. (2012). Hardware-software system for simulating and analyzing earthquakes applied to civil structures. *Natural Hazards & Earth System Sciences*, *12*(1), 61–73.
99. Hsu, T. Y., Huang, S. K., Lu, K. C., Loh, C. H., Wang, Y., & Lynch, J. P. (2011). On-line structural damage localization and quantification using wireless sensors. *Smart Materials and Structures*, *20*(10), 105025.
100. García-Palencia, A. J., & Santini-Bell, E. (2013). A two-step model updating algorithm for parameter identification of linear elastic damped structures. *Computer-Aided Civil and Infrastructure Engineering*, *28*(7), 509–521.
101. Hu, X., Wang, B., & Ji, H. (2013). A wireless sensor network-based structural health monitoring system for highway bridges. *Computer-Aided Civil and Infrastructure Engineering*, *28*(3), 193–209.

102. Torbol, M., Gomez, H., & Feng, M. (2013). Fragility analysis of highway bridges based on long-term monitoring data. *Computer-Aided Civil and Infrastructure Engineering*, *28*(3), 178–192.
103. Dinh, V. N., Basu, B., & Brinkgreve, R. B. (2014). Wavelet-based evolutionary response of multispan structures including wave-passage and site-response effects. *Journal of Engineering Mechanics*, *140*(8), 04014056.
104. Qiao, L., Esmaeily, A., & Melhem, H. G. (2012). Signal pattern recognition for damage diagnosis in structures. *Computer-Aided Civil and Infrastructure Engineering*, *27*(9), 699–710.
105. Tang, J. P., Chiou, D. J., Chen, C. W., Chiang, W. L., Hsu, W. K., Chen, C. Y., & Liu, T. Y. (2011). RETRACTED: A case study of damage detection in benchmark buildings using a Hilbert-Huang transform-based method. *Journal of Vibration and Control*, *17*(4), 623–636.
106. El-Shafie, A., Noureldin, A., McGaughey, D., & Hussain, A. (2012). Fast orthogonal search (FOS) versus fast Fourier transform (FFT) as spectral model estimations techniques applied for structural health monitoring (SHM). *Structural and Multidisciplinary Optimization*, *45*(4), 503–513.
107. Philibert, M., Soutis, C., Gresil, M., & Yao, K. (2018). Damage detection in a composite T-joint using guided lamb waves. *Aerospace*, *5*(2), 40.
108. Lynch, J. P., Sundararajan, A., Law, K. H., Kiremidjian, A. S., Kenny, T., & Carryer, E. (2003). Embedment of structural monitoring algorithms in a wireless sensing unit. *Structural Engineering and Mechanics*, *15*(3), 285–297.
109. Zhang, F. L., Yang, Y. P., Xiong, H. B., Yang, J. H., & Yu, Z. (2019). Structural health monitoring of a 250-m super-tall building and operational modal analysis using the fast Bayesian FFT method. *Structural Control and Health Monitoring*, *26*(8), e2383.
110. Loewke, K., Meyer, D., Starr, A., & Nemat-Nasser, S. (2005, May). Structural health monitoring using FFT. In *Smart Structures and Materials 2005: Sensors and Smart Structures Technologies for Civil, Mechanical, and Aerospace Systems* (Vol. 5765, pp. 931–935). International Society for Optics and Photonics. doi: 10.1117/12.598827.
111. Alsaadi, A., Shi, Y., & Jia, Y. (2020). Delamination detection via reconstructed frequency response function of composite structures. In M. Wahab (ed.), *Proceedings of the 13th International Conference on Damage Assessment of Structures* (pp. 837–843). Singapore: Springer.
112. Zenzen, R., Belaidi, I., Khatir, S., & Wahab, M. A. (2018). A damage identification technique for beam-like and truss structures based on FRF and Bat Algorithm. *Comptes Rendus Mécanique*, *346*(12), 1253–1266.
113. Lin, J. H., & Loh, C. H. (2017, April). Structural damage detection using high dimension data reduction and visualization techniques. In *Sensors and Smart Structures Technologies for Civil, Mechanical, and Aerospace Systems 2017* (Vol. 10168, p. 101682U). International Society for Optics and Photonics. doi: 10.1117/12.2257896.
114. Pu, Q., Hong, Y., Chen, L., Yang, S., & Xu, X. (2019). Model updating-based damage detection of a concrete beam utilizing experimental damped frequency response functions. *Advances in Structural Engineering*, *22*(4), 935–947.
115. Zang, C., Friswell, M. I., & Imregun, M. (2003). Structural health monitoring and damage assessment using measured FRFs from multiple sensors, part I: The indicator of correlation criteria. In *Key Engineering Materials* (Vol. 245, pp. 131–140). Trans Tech Publications Ltd. 10.4028/www.scientific.net/KEM.245-246.131.

116. Sampaio, R. P. C., Maia, N. M. M., & Silva, J. M. M. (1999). Damage detection using the frequency-response-function curvature method. *Journal of Sound and Vibration*, *226*(5), 1029–1042.
117. Sampaio, R. P. C., Maia, N. M. M., & Silva, J. M. M. (2003). The frequency domain assurance criterion as a tool for damage detection. In *Key Engineering Materials* (Vol. 245, pp. 69–76). Trans Tech Publications Ltd. 10.4028/www.scientific.net/KEM.245-246.69.
118. Mondal, S., Mondal, B., Bhutia, A., & Chakraborty, S. (2015). Damage detection in beams using frequency response function curvatures near resonating frequencies. In V. Matsagar (ed.) *Advances in Structural Engineering* (pp. 1563–1573). New Delhi: Springer.
119. Reddy, D. M., & Swarnamani, S. (2012). Application of the FRF curvature energy damage detection method to plate like structures. *World Journal of Modelling and Simulation*, *8*(2), 147–153.
120. Kim, B. H., Park, T., & Voyiadjis, G. Z. (2006). Damage estimation on beam-like structures using the multi-resolution analysis. *International Journal of Solids and Structures*, *43*(14–15), 4238–4257.
121. Shi, J. Y., Spencer Jr., B. F., & Chen, S. S. (2018). Damage detection in shear buildings using different estimated curvature. *Structural Control and Health Monitoring*, *25*(1), e2050.
122. Liu, X., Lieven, N. A. J., & Escamilla-Ambrosio, P. J. (2009). Frequency response function shape-based methods for structural damage localisation. *Mechanical Systems and Signal Processing*, *23*(4), 1243–1259.
123. Lin, R. M., & Ewins, D. J. (1994). Analytical model improvement using frequency response functions. *Mechanical Systems and Signal Processing*, *8*(4), 437–458.
124. Lin, R. M., & Zhu, J. (2006). Model updating of damped structures using FRF data. *Mechanical Systems and Signal Processing*, *20*(8), 2200–2218.
125. Wang, Z., Lin, R. M., & Lim, M. K. (1997). Structural damage detection using measured FRF data. *Computer Methods in Applied Mechanics and Engineering*, *147*(1–2), 187–197.
126. Esfandiari, A., Rahai, A., Sanayei, M., & Bakhtiari-Nejad, F. (2016). Model updating of a concrete beam with extensive distributed damage using experimental frequency response function. *Journal of Bridge Engineering*, *21*(4), 04015081.
127. Sipple, J. D., & Sanayei, M. (2014). Finite element model updating using frequency response functions and numerical sensitivities. *Structural Control and Health Monitoring*, *21*(5), 784–802.
128. Mohan, S. C., Maiti, D. K., & Maity, D. (2013). Structural damage assessment using FRF employing particle swarm optimization. *Applied Mathematics and Computation*, *219*(20), 10387–10400.
129. Hong, Y., Liu, X., Dong, X., Wang, Y., & Pu, Q. (2016, April). Experimental model updating using frequency response functions. In *Sensors and Smart Structures Technologies for Civil, Mechanical, and Aerospace Systems 2016* (Vol. 9803, p. 980325). International Society for Optics and Photonics. doi: 10.1117/12.2219364.
130. Hong, Y., Pu, Q., Wang, Y., Chen, L., Gou, H., & Li, X. (2018). Model-updating with experimental frequency response function considering general damping. *Advances in Structural Engineering*, *21*(1), 82–92.
131. Valdés-González, J., De-la-Colina, J., & González-Pérez, C. A. (2015). Experiments for seismic damage detection of a RC frame using ambient and forced vibration records. *Structural Control and Health Monitoring*, *22*(2), 330–346.

132. de Castro, B. A., Baptista[1], F. G., & Ciampa, F. (2018, January). Impedance-based structural health monitoring under low signal-to-noise ratio conditions. In *9th European Workshop on Structural Health Monitoring, EWSHM 2018*, Manchester.
133. Padil, K. H., Bakhary, N., Abdulkareem, M., Li, J., & Hao, H. (2020). Non-probabilistic method to consider uncertainties in frequency response function for vibration-based damage detection using Artificial Neural Network. *Journal of Sound and Vibration*, *467*, 115069.
134. Dackermann, U., Li, J., & Samali, B. (2013). Identification of member connectivity and mass changes on a two-storey framed structure using frequency response functions and artificial neural networks. *Journal of Sound and Vibration*, *332*, 3636–3653.
135. Nguyen, V., Dackermann, U., Li, J., Alamdari, M. M., Mustapha, S., Runcie, P., & Ye, L. (2015). Damage identification of a concrete arch beam based on frequency response functions and artificial neural networks. *Electronic Journal of Structural Engineering*, *14*, 75–84.
136. Khoshnoudian, F., Talaei, S., & Fallahian, M. (2016). Structural damage detection using FRF Data, 2D-PCA, artificial neural networks and imperialist competitive algorithm simultaneously. *International Journal of Structural Stability and Dynamics*, *17*, 1750073.
137. Khoshnoudian, F., & Talaei, S. (2016). A new damage index using FRF data, 2D-PCA method and pattern recognition techniques. *International Journal of Structural Stability and Dynamics*, *17*, 1750090.
138. Wang, J., Wang, C., & Zhao, J. (2017). Frequency response function-based model updating using Kriging model. *Mechanical Systems and Signal Processing*, *87*, 218–228.
139. Fallahian, M., Khoshnoudian, F., & Talaei, S. (2018). Application of couple sparse coding ensemble on structural damage detection. *Smart Structures and Systems*, *21*(1), 001–14.
140. Ip, C. U., & Vickery, C. A. (1987). Dynamic stress at critical locations of a structure as a criterion for mathematical model modification. In *in Shock and Vibration Information Center The Shock and Vibration Bulletin. Part 4: Structural Dynamics and Modal Test and Analysis* (pp. 29–46) (SEE N 88-17062 09-39) (Vol. 1987).
141. Esfandiari, A., Sanayei, M., Bakhtiari-Nejad, F., & Rahai, A. (2010). Finite element model updating using frequency response function of incomplete strain data. *AIAA Journal*, *48*(7), 1420–1433.
142. Guo, N., Yang, Z., Jia, Y., & Wang, L. (2016). Model updating using correlation analysis of strain frequency response function. *Mechanical Systems and Signal Processing*, *70*, 284–299.
143. Cornwell, P., Doebling, S. W., & Farrar, C. R. (1999). Application of the strain energy damage detection method to plate-like structures. *Journal of Sound and Vibration*, *224*(2), 359–374.
144. Yam, L. H., Li, Y. Y., & Wong, W. O. (2002). Sensitivity studies of parameters for damage detection of plate-like structures using static and dynamic approaches. *Engineering Structures*, *24*(11), 1465–1475.
145. Lee, J., & Kim, S. (2007). Structural damage detection in the frequency domain using neural networks. *Journal of Intelligent Material Systems and Structures*, *18*(8), 785–792.

146. Park, H. J., Koo, K. Y., & Yun, C. B. (2007). Modal flexibility-based damage detection technique of steel beam by dynamic strain measurements using FBG sensors. *Steel Structures*, *7*, 11–18.
147. Wu, Z., & Li, S. (2007). Two-level damage detection strategy based on modal parameters from distributed dynamic macro-strain measurements. *Journal of Intelligent Material Systems and Structures*, *18*(7), 667–676.
148. Kesavan, A., John, S., & Herszberg, I. (2008). Strain-based structural health monitoring of complex composite structures. *Structural Health Monitoring*, *7*(3), 203–213.
149. Katsikeros, C. E., & Labeas, G. N. (2009). Development and validation of a strain-based structural health monitoring system. *Mechanical Systems and Signal Processing*, *23*(2), 372–383.
150. Li, Y. Y. (2010). Hypersensitivity of strain-based indicators for structural damage identification: A review. *Mechanical Systems and Signal Processing*, *24*(3), 653–664.
151. Liu, M., Ke, M., Zhou, Z., & Tan, Y. (2011, August). Strain response frequency function-based mechanical damage indentification by fiber bragg grating sensors. In *2011 Second International Conference on Digital Manufacturing & Automation*, Zhangjiajie (pp. 1117–1120). IEEE.
152. Loutas, T. H., Panopoulou, A., Roulias, D., & Kostopoulos, V. (2012). Intelligent health monitoring of aerospace composite structures based on dynamic strain measurements. *Expert Systems with Applications*, *39*(9), 8412–8422.
153. Lee, E. T., Rahmatalla, S., & Eun, H. C. (2013). Damage detection by mixed measurements using accelerometers and strain gages. *Smart Materials and Structures*, *22*(7), 075014.
154. Lee, E. T., & Eun, H. C. (2014). Damage detection of beam structure using response data measured by strain gages. *Journal of Vibroengineering*, *16*(1), 147–155.
155. Zhang, J., Guo, S. L., Wu, Z. S., & Zhang, Q. Q. (2015). Structural identification and damage detection through long-gauge strain measurements. *Engineering Structures*, *99*, 173–183.
156. Cheng, L., & Cigada, A. (2017). Experimental strain modal analysis for beam-like structure by using distributed fiber optics and its damage detection. *Measurement Science and Technology*, *28*(7), 074001.
157. Shadan, F., Khoshnoudian, F., & Esfandiari, A. (2018). Structural damage identification based on strain frequency response functions. *International Journal of Structural Stability and Dynamics*, *18*(12), 1850159.
158. Brincker R, Zhang L, Andersen P. (2000). Modal identification from ambient responses using frequency domain decomposition. In: *Proceedings of the 18th IMAC*, San Antonio, TX.
159. Brincker, R., Ventura, C. E., & Andersen, P. (2001). Damping estimation by frequency domain decomposition. In *Proceedings of IMAC 19: A Conference on Structural Dynamics*: February 5–8, 2001, Hyatt Orlando, Kissimmee, FL, 2001 (pp. 698–703). Society for Experimental Mechanics.
160. Wang, T., & Zhang, L. M. (2006). Frequency and spatial domain decomposition for operational modal analysi2sand its application. *Acta Aeronautica et Astronautica Sinica*, *27*, pp. 62–66.
161. Hui, L., Huang, Ch, & Xun, Y. (2015). Modal test and analysis of the tied arch bridge by using EFDD method. *Highway Engineering*, *40*, pp. 129–132.

162. Altunisik, A. C., & Bayraktar, A. (2012). Baris Sevim Operational modal analysis of a scaled bridge model using EFDD and SSI methods. *Indian Journal of Engineering and Materials Sciences*, *19*, 320–330.
163. Foti, D., Gattulli, V., Potenza, F. (2014). Output-only identification and model updating by dynamic testing in unfavorable conditions of a seismically damaged building. *Computer-Aided Civil and Infrastructure Engineering*, *29*, 659–675.
164. Wu, J., Li, H., Ye, F., & Ma, K. (2019). Damage identification of bridge structure based on frequency domain decomposition and strain mode. *Journal of Vibroengineering*, *21*(8), 2096–2105.
165. Makki Alamdari, M., Anaissi, A., Khoa, N. L., & Mustapha, S. (2019). Frequency domain decomposition-based multisensor data fusion for assessment of progressive damage in structures. *Structural Control and Health Monitoring*, *26*(2), e2299.
166. Abdelghani, M., Ghalishooyan, M., & Shooshtari, A. (2017). A comparative assessment of in-operation modal analysis and frequency domain decomposition algorithm using simulated data. In T. Fakhfakh, F. Chaari, L. Walha, M. Abdennadher, M. Abbes, M. Haddar (eds.), *Advances in Acoustics and Vibration* (pp. 215–221). Cham: Springer.
167. Liu, G., & Venkatasubramanian, V. (2008, May). Oscillation monitoring from ambient PMU measurements by frequency domain decomposition. In *2008 IEEE International Symposium on Circuits and Systems*, Seattle, WA (pp. 2821–2824). IEEE.
168. Pioldi, F., Ferrari, R., & Rizzi, E. (2017). Earthquake structural modal estimates of multi-storey frames by a refined Frequency Domain Decomposition algorithm. *Journal of Vibration and Control*, *23*(13), 2037–2063.
169. Górski, P. (2017). Dynamic characteristic of tall industrial chimney estimated from GPS measurement and frequency domain decomposition. *Engineering Structures*, *148*, 277–292.
170. Magalhães, F., & Cunha, A. (2011). Explaining operational modal analysis with data from an arch bridge. *Mechanical Systems and Signal Processing*, *25*(5), 1431–1450.
171. Brincker, R., Zhang, L., & Andersen, P. (2000, September). Output-only modal analysis by frequency domain decomposition. In *Proceedings of the ISMA25 Noise and Vibration Engineering*, Leuven (Vol. 11, pp. 717–723).
172. Malekjafarian, A., & OBrien, E. J. (2014). Identification of bridge mode shapes using short time frequency domain decomposition of the responses measured in a passing vehicle. *Engineering Structures*, *81*, 386–397.
173. Stoica, P., & Moses, R. L. (1997). *Introduction to Spectral Analysis*. Englewood Cliffs, NJ: Prentice-Hall.
174. Nagata, Y., Iwasaki, S., Hariyama, T., Fujioka, T., Obara, T., & Wakatake, T. (2009). Binaural localization based on weighted wiener gain improved by incremental source attenuation. *IEEE Transaction on Audio Speech and Language Processing*, *17*(1), 52–65.
175. Garcia-Perez, A., Romero-Troncoso, R. J., Cabal-Yepez, E., Osornio- Rios, R. A., Lucio-Martinez, J. A. (2012) Application of highresolution spectral analysis for identifying faults in induction motors by means of sound. *Journal of Vibration and Control* *18*(11), 1585–1594.

176. Adeli, H., & Jiang, X. (2006) Dynamic fuzzy wavelet neural network model for structural system identification. *Journal of Structural Engineering ASCE 132*(1), 102–111.
177. Jiang, X., Adeli, H. (2007) Pseudospectra, MUSIC, and dynamic wavelet neural network for damage detection of highrise buildings. *International Journal for Numerical Methods in Engineering, 71*(5), 606–629.
178. Osornio-Rios, R. A., Amezquita-Sanchez, J. P., Romero-Troncoso, R. J., Garcia-Perez, A. (2012). MUSIC-ANN analysis for locating structural damages in a truss-type structure by means of vibrations. *Computer-Aided Civil and Infrastructure Engineering, 27*(9), 687–698.
179. Garcia-Perez, A., Romero-Troncoso, R. J., Cabal-Yepez, E., Osornio-Rios, R. A., Rangel-Magdaleno, J. J., Miranda, H. (2011). Startup current analysis of incipient broken rotor bar in induction motors using high-resolution spectral analysis. In: *Proceedings of IEEE International Symposium on Diagnostics for Electric Machines, Power Electronics and Drives*, Bologna, Italy, September 5–8, 2011 (pp. 657–663).
180. Amezquita-Sanchez, J. P., Garcia-Perez, A., Romero-Troncoso, R. J., Osornio-Rios, R. A., Herrera-Ruiz, G. (2013) High-resolution spectral-analysis for identifying the natural modes of a truss-type structure by means of vibrations. *Journal of Vibration and Control, 19*, 2347–2356.
181. Bao C, Hao H, Li Z, Zhu X (2009) Time-varying system identification using a newly improved HHT algorithm. *Computers & Structures, 87*(23-24), 611–1623.
182. Perez, A. G., Troncoso, R. J. R., Yepez, E. C., Rios, R. A. O., & Martinez, J. A. L. (2012). Application of highresolution spectral analysis for identifying faults in induction motors by means of sound. *Journal of Vibration and Control, 18*, 1585–1594.
183. Martinez, D. C., Rios, R. O., Troncoso, R. J. R., & Perez, A. G. (2015). Fused empirical mode decomposition and MUSIC algorithms for detecting multiple combined faults in induction motors. *Journal of Applied Research and Technology, 13*, 160–167.
184. Camarena-Martinez, D., Amezquita-Sanchez, J. P., Valtierra-Rodriguez, M., Romero-Troncoso, R. J., Osornio-Rios, R. A., & Garcia-Perez, A. (2014). EEMD-MUSIC-based analysis for natural frequencies identification of structures using artificial and natural excitations. *The Scientific World Journal*, 2014, 587671.
185. Zhong, Y., Yuan, S., & Qiu, L. (2014). Multiple damage detection on aircraft composite structures using near-field MUSIC algorithm. *Sensors and Actuators A: Physical, 214*, 234–244.
186. Zuo, H., Yang, Z., Xu, C., Tian, S., & Chen, X. (2018). Damage identification for plate-like structures using ultrasonic guided wave based on improved MUSIC method. *Composite Structures, 203*, 164–171.
187. Perez-Ramirez, C. A., Machorro-Lopez, J. M., Valtierra-Rodriguez, M., Amezquita-Sanchez, J. P., Garcia-Perez, A., Camarena-Martinez, D., & Romero-Troncoso, R. D. J. (2020). Location of multiple damage types in a truss-type structure using multiple signal classification method and vibration signals. *Mathematics, 8*(6), 932.

188. Gkoktsi, K., & Giaralis, A. (2020). A compressive MUSIC spectral approach for identification of closely-spaced structural natural frequencies and post-earthquake damage detection. *Probabilistic Engineering Mechanics*, *60*, 103030.
189. Bao, Q., Yuan, S., Wang, Y., & Qiu, L. (2019). Anisotropy compensated MUSIC algorithm based composite structure damage imaging method. *Composite Structures*, *214*, 293–303.
190. Elbouchikhi, E., Choqueuse, V., & Benbouzid, M. (2016). Induction machine bearing faults detection based on a multi-dimensional MUSIC algorithm and maximum likelihood estimation. *ISA Transactions*, *63*, 413–424.
191. Bao, Q., Yuan, S., & Guo, F. (2020). A new synthesis aperture-MUSIC algorithm for damage diagnosis on complex aircraft structures. *Mechanical Systems and Signal Processing*, *136*, 106491.
192. Fu, T., Wang, Y., Qiu, L., & Tian, X. (2020). Sector piezoelectric sensor array transmitter beamforming MUSIC algorithm based structure damage imaging method. *Sensors*, *20*(5), 1265.
193. Yuan, S., Bao, Q., Qiu, L., & Zhong, Y. (2015). A single frequency component-based re-estimated MUSIC algorithm for impact localization on complex composite structures. *Smart Materials and Structures*, *24*(10), 105021.
194. Dimopoulos, V., Becht, P., Janssens, D., Deckers, E., & Desmet, W. (2019). Efficient TR-MUSIC damage detection in composites with a limited number of sensors. *Proceeding of NDT Aerospace*, 2019, 25.
195. He, J., & Yuan, F. G. (2016). Lamb wave-based subwavelength damage imaging using the DORT-MUSIC technique in metallic plates. *Structural Health Monitoring*, *15*(1), 65–80.
196. Zhang, C., Mousavi, A. A., Masri, S. F., Gholipour, G., Yan, K., & Li, X. (2022). Vibration feature extraction using signal processing techniques for structural health monitoring: A review. *Mechanical Systems and Signal Processing*, *177*, 109175.
197. Amini, F., Hazaveh, N. K., & Rad, A. A. (2013). Wavelet PSO-based LQR algorithm for optimal structural control using active tuned mass dampers. *Computer-Aided Civil and Infrastructure Engineering*, *28*(7), 542–557.
198. Katicha, S. W., Flintsch, G., Bryce, J., & Ferne, B. (2014). Wavelet denoising of TSD deflection slope measurements for improved pavement structural evaluation. *Computer-Aided Civil and Infrastructure Engineering*, *29*(6), 399–415.
199. Amini, F., & Samani, M. Z. (2014). A wavelet-based adaptive pole assignment method for structural control. *Computer-Aided Civil and Infrastructure Engineering*, *29*(6), 464–477.
200. Nigro, M. B., Pakzad, S. N., & Dorvash, S. (2014). Localized structural damage detection: A change point analysis. *Computer-Aided Civil and Infrastructure Engineering*, *29*(6), 416–432.
201. Khalid, M., Yusof, R., Joshani, M., Selamat, H., & Joshani, M. (2014). Nonlinear identification of a magneto-rheological damper based on dynamic neural networks. *Computer-Aided Civil and Infrastructure Engineering*, *29*(3), 221–233.
202. Peng, Z. K., & Chu, F. L. (2004). Application of the wavelet transform in machine condition monitoring and fault diagnostics: A review with bibliography. *Mechanical Systems and Signal Processing*, *18*(2), 199–221.

203. Nguyen, H. N., Kim, J., & Kim, J. M. (2018). Optimal sub-band analysis based on the envelope power Spectrum for effective fault detection in bearing under variable, low speeds. *Sensors*, *18*(5), 1389.
204. Baek, W., Baek, S., & Kim, D. Y. (2018). Characterization of system status signals for multivariate time series discretization based on frequency and amplitude variation. *Sensors*, *18*(1), 154.
205. Yan, R., Gao, R. X., & Chen, X. (2014). Wavelets for fault diagnosis of rotary machines: A review with applications. *Signal Processing*, *96*, 1–15.
206. Feng, Z., Liang, M., & Chu, F. (2013). Recent advances in time-frequency analysis methods for machinery fault diagnosis: A review with application examples. *Mechanical Systems and Signal Processing*, *38*(1), 165–205.
207. Lei, Y., Lin, J., He, Z., & Zuo, M. J. (2013). A review on empirical mode decomposition in fault diagnosis of rotating machinery. *Mechanical Systems and Signal Processing*, *35*(1–2), 108–126.
208. Gabor, D. (1946). Theory of communication. Part 1: The analysis of information. *Journal of the Institution of Electrical Engineers-Part III: Radio and Communication Engineering*, *93*(26), 429–441.
209. Cohen, L. (1995). *Time-Frequency Analysis* (Vol. 778). Englewood Cliffs, NJ: Prentice Hall.
210. Yinfeng, D., Yingmin, L., Mingkui, X., & Ming, L. (2008). Analysis of earthquake ground motions using an improved Hilbert-Huang transform. *Soil Dynamics and Earthquake Engineering*, *28*(1), 7–19.
211. Amezquita-Sanchez, J. P., Garcia-Perez, A., Romero-Troncoso, R. J., Osornio-Rios, R. A., & Herrera-Ruiz, G. (2013). High-resolution spectral-analysis for identifying the natural modes of a truss-type structure by means of vibrations. *Journal of Vibration and Control*, *19*(16), 2347–2356.
212. Yesilyurt, I., & Gursoy, H. (2015). Estimation of elastic and modal parameters in composites using vibration analysis. *Journal of Vibration and Control*, *21*(3), 509–524.
213. Dolce, M., & Cardone, D. (2006). Theoretical and experimental studies for the application of shape memory alloys in civil engineering, 302–311. doi: 10.1115/1.2203106.
214. Cocconcelli, M., Zimroz, R., Rubini, R., & Bartelmus, W. (2012). STFT based approach for ball bearing fault detection in a varying speed motor. In T. Fakhfakh, W. Bartelmus, F. Chaari, R. Zimroz, & M. Haddar (eds.), *Condition Monitoring of Machinery in Non-Stationary Operations* (pp. 41–50). Heidelberg: Springer.
215. Nagarajaiah, S., & Basu, B. (2009). Output only modal identification and structural damage detection using time frequency & wavelet techniques. *Earthquake Engineering and Engineering Vibration*, *8*(4), 583–605.
216. Nagarajaiah, S., Nadathur, V., & Sahasrabudhe, S. (1999). Variable stiffness and instantaneous frequency. In *Proceedings of Structures Congress*, ASCE, New Orleans, 858–861.
217. Nagarajaiah, S., & Varadarajan, N. (1997) Semi-active control of smart tuned mass damper using empirical mode decomposition and Hilbert transform algorithm. *Engineering Geology and the Environment*, *205*, 800–841.
218. Nagarajaiah, S., & Sonmez, E. (2007). Structures with semiactive variable stiffness single/multiple tuned mass dampers. *Journal of Structural Engineering*, *133*(1), 67–77.

219. Nagarajaiah, S. (2009). Adaptive passive, semiactive, smart tuned mass dampers: Identification and control using empirical mode decomposition, Hilbert transform, and short-term Fourier transform. *Structural Control and Health Monitoring: The Official Journal of the International Association for Structural Control and Monitoring and of the European Association for the Control of Structures*, *16*(7–8), 800–841.
220. Narasimhan, S., & Nagarajaiah, S. (2005). A STFT semiactive controller for base isolated buildings with variable stiffness isolation systems. *Engineering Structures*, *27*(4), 514–523.
221. Varadarajan, N., & Nagarajaiah, S. (2004). Wind response control of building with variable stiffness tuned mass damper using empirical mode decomposition/Hilbert transform. *Journal of Engineering Mechanics*, *130*(4), 451–458.
222. Kim, B. S., Lee, S. H., Lee, M. G., Ni, J., Song, J. Y., & Lee, C. W. (2007). A comparative study on damage detection in speed-up and coast-down process of grinding spindle-typed rotor-bearing system. *Journal of Materials Processing Technology*, *187*, 30–36.
223. Fitzgerald, B., Arrigan, J., & Basu, B. (2010, July). Damage detection in wind turbine blades using time-frequency analysis of vibration signals. In *The 2010 International Joint Conference on Neural Networks (IJCNN)*, Barcelona (pp. 1–5). IEEE.
224. Mousavi, A. A., Zhang, C., Masri, S. F., & Gholipour, G. (2020). Structural damage localization and quantification based on a CEEMDAN Hilbert transform neural network approach: A model steel truss bridge case study. *Sensors*, *20*(5), 1271.
225. Gurley, K., & Kareem, A. (1999). Applications of wavelet transforms in earthquake, wind and ocean engineering. *Engineering Structures*, *21*(2), 149–167.
226. Xu, Y. L., & Chen, B. (2008). Integrated vibration control and health monitoring of building structures using semi-active friction dampers: Part I-methodology. *Engineering Structures*, *30*(7), 1789–1801.
227. Chen, B., & Xu, Y. L. (2008). Integrated vibration control and health monitoring of building structures using semi-active friction dampers: Part II-numerical investigation. *Engineering Structures*, *30*(3), 573–587.
228. Ewins, D. J. (2009). *Modal Testing: Theory, Practice and Application*. Hoboken, NJ: John Wiley & Sons.
229. Li, H., Yi, T., Gu, M., & Huo, L. (2009). Evaluation of earthquake-induced structural damages by wavelet transform. *Progress in Natural Science*, *19*(4), 461–470.
230. Yi, T. H., Li, H. N., & Zhao, X. Y. (2012). Noise smoothing for structural vibration test signals using an improved wavelet thresholding technique. *Sensors*, *12*(8), 11205–11220.
231. Yi, T. H., Li, H. N., & Gu, M. (2013). Wavelet based multi-step filtering method for bridge health monitoring using GPS and accelerometer. *Smart Structures and Systems*, *11*(4), 331–348.
232. Huang, N. E., Long, S. R., & Shen, Z. (1996). The mechanism for frequency downshift in nonlinear wave evolution. In *Advances in Applied Mechanics* (Vol. 32, pp. 59–117C). Elsevier. 10.1016/S0065-2156(08)70076-0.
233. Huang, N. E., Long, S. R., & Shen, Z. (1996). The mechanism for frequency downshift in nonlinear wave evolution. In *Advances in Applied Mechanics* (Vol. 32, pp. 59–117C). Elsevier. 10.1016/S0065-2156(08)70076-0.

234. Huang, N. E., Shen, Z., & Long, S. R. (1999). A new view of nonlinear water waves: The Hilbert spectrum. *Annual Review of Fluid Mechanics*, *31*(1), 417–457.
235. Huang, N. E., & Attoh-Okine, N. O. (2005). *The Hilbert-Huang Transform in Engineering*. Boca Raton, FL: CRC Press.
236. Vincent, H. T., Hu, S. L. J., & Hou, Z. (1999). Damage detection using empirical mode decomposition method and a comparison with wavelet analysis. In *Proceedings of the 2nd International Workshop on Structural Health Monitoring* (pp. 891–900). Standford, CA: Stanford University.
237. Yang, J. N., & Lei, Y. (1999, October). Identification of natural frequencies and damping ratios of linear structures via Hilbert transform and empirical mode decomposition. In *Proceedings of the International Conference on Intelligent Systems and Control* (pp. 310–315). Anaheim, CA: IASTED/Acta Press.
238. Yang, J. N., & Lei, Y. (2000). System Identification of linear structures using Hilbert transform and empirical mode decomposition, # 256. In *Proceedings of IMAC-XVIII: A Conference on Structural Dynamics*, Irvine, CA (Vol. 4062, p. 213).
239. Yang, J. N., & Lei, Y. (2001). Damage identification of civil engineering structures using Hilbert-Huang transform. In *Proceedings of the 3rd International Workshop on Structural Health Monitoring*, New York (pp. 544–553).
240. Rezaei, D., & Taheri, F. (2011). Damage identification in beams using empirical mode decomposition. *Structural Health Monitoring*, *10*(3), 261–274.
241. Dong, Y., Li, Y., & Lai, M. (2010). Structural damage detection using empirical-mode decomposition and vector autoregressive moving average model. *Soil Dynamics and Earthquake Engineering*, *30*(3), 133–145.
242. Cheng-Zhong, Q., & Xu-Wei, L. (2012). Damage identification for transmission towers based on HHT. *Energy Procedia*, *17*, 1390–1394.
243. Xu, Y. L., & Chen, J. (2004). Structural damage detection using empirical mode decomposition: Experimental investigation. *Journal of Engineering Mechanics*, *130*(11), 1279–1288.
244. Sarmadi, H., Entezami, A., & Daneshvar Khorram, M. (2020). Energy-based damage localization under ambient vibration and non-stationary signals by ensemble empirical mode decomposition and Mahalanobis-squared distance. *Journal of Vibration and Control*, *26*(11–12), 1012–1027.
245. Entezami, A., & Shariatmadar, H. (2019). Structural health monitoring by a new hybrid feature extraction and dynamic time warping methods under ambient vibration and non-stationary signals. *Measurement*, *134*, 548–568.
246. Fakih, M. A., Mustapha, S., Tarraf, J., Ayoub, G., & Hamade, R. (2018). Detection and assessment of flaws in friction stir welded joints using ultrasonic guided waves: Experimental and finite element analysis. *Mechanical Systems and Signal Processing*, *101*, 516–534.
247. Pines, D., & Salvino, L. (2006). Structural health monitoring using empirical mode decomposition and the Hilbert phase. *Journal of Sound and Vibration*, *294*(1–2), 97–124.
248. Chen, B., Zhao, S. L., & Li, P. Y. (2014). Application of Hilbert-Huang transform in structural health monitoring: A state-of-the-art review. *Mathematical Problems in Engineering*, 2014, 1–22.
249. Xu, Y. L., & Chen, J. (2003). Empirical mode decomposition for structural damage detection. In *International Conference on Inspection, Appraisal, Repairs and Maintenance of Structures*, Hong Kong.

250. Bao, C., Hao, H., Li, Z. X., & Zhu, X. (2009). Time-varying system identification using a newly improved HHT algorithm. *Computers & Structures*, *87*(23–24), 1611–1623.
251. He, X. H., Hua, X. G., Chen, Z. Q., & Huang, F. L. (2011). EMD-based random decrement technique for modal parameter identification of an existing railway bridge. *Engineering Structures*, *33*(4), 1348–1356.
252. Pavlopoulou, S., Staszewski, W. J., & Soutis, C. (2013). Evaluation of instantaneous characteristics of guided ultrasonic waves for structural quality and health monitoring. *Structural Control and Health Monitoring*, *20*(6), 937–955.
253. Ghazali, M. F., Staszewski, W. J., Shucksmith, J. D., Boxall, J. B., & Beck, S. B. M. (2011). Instantaneous phase and frequency for the detection of leaks and features in a pipeline system. *Structural Health Monitoring*, *10*(4), 351–360.
254. Esmaeel, R. A., & Taheri, F. (2012). Delamination detection in laminated composite beams using the empirical mode decomposition energy damage index. *Composite Structures*, *94*(5), 1515–1523.
255. Alvanitopoulos, P. F., Andreadis, I., & Elenas, A. (2009). Interdependence between damage indices and ground motion parameters based on Hilbert-Huang transform. *Measurement Science and Technology*, *21*(2), 025101.
256. Wei, Y. C., Lee, C. J., Hung, W. Y., & Chen, H. T. (2010). Application of Hilbert-Huang transform to characterize soil liquefaction and quay wall seismic responses modeled in centrifuge shaking-table tests. *Soil Dynamics and Earthquake Engineering*, *30*(7), 614–629.
257. Shi, W., Shan, J., & Lu, X. (2012). Modal identification of Shanghai World Financial Center both from free and ambient vibration response. *Engineering Structures*, *36*, 14–26.
258. Garcia-Perez, A., Amezquita-Sanchez, J. P., Dominguez-Gonzalez, A., Sedaghati, R., Osornio-Rios, R., & Romero-Troncoso, R. J. (2013). Fused empirical mode decomposition and wavelets for locating combined damage in a truss-type structure through vibration analysis. *Journal of Zhejiang University SCIENCE A*, *14*(9), 615–630.
259. Chiou, D. J., Hsu, W. K., Chen, C. W., Hsieh, C. M., Tang, J. P., & Chiang, W. L. (2011). Applications of Hilbert-Huang transform to structural damage detection. *Structural Engineering and Mechanics: An International Journal*, *39*(1), 1–20.
260. Lin, L., & Chu, F. (2011). Feature extraction of AE characteristics in offshore structure model using Hilbert-Huang transform. *Measurement*, *44*(1), 46–54.
261. Hamdi, S. E., Le Duff, A., Simon, L., Plantier, G., Sourice, A., & Feuilloy, M. (2013). Acoustic emission pattern recognition approach based on Hilbert-Huang transform for structural health monitoring in polymer-composite materials. *Applied Acoustics*, *74*(5), 746–757.
262. Lin, C. C., Liu, P. L., & Yeh, P. L. (2009). Application of empirical mode decomposition in the impact-echo test. *NDT & E International*, *42*(7), 589–598.
263. Yadav, S. K., Banerjee, S., & Kundu, T. (2011). Effective damage sensitive feature extraction methods for crack detection using flaw scattered ultrasonic wave field signal. In *8th International Workshop on Structural Health Monitoring 2011: Condition-Based Maintenance and Intelligent Structures*, Stanford, CA (pp. 167–174).

264. Meredith, J., González, A., & Hester, D. (2012). Empirical mode decomposition of the acceleration response of a prismatic beam subject to a moving load to identify multiple damage locations. *Shock and Vibration*, *19*(5), 845–856.
265. Roveri, N., & Carcaterra, A. (2012). Damage detection in structures under traveling loads by Hilbert-Huang transform. *Mechanical Systems and Signal Processing*, *28*, 128–144.
266. Yu, D. J., & Ren, W. X. (2005). EMD-based stochastic subspace identification of structures from operational vibration measurements. *Engineering Structures*, *27*(12), 1741–1751.
267. Ma, L., Liu, J. X., Han, W. S., & Ji, B. H. (2010). Time-frequency analysis for nonlinear buffeting response of a long-span bridge based on HHT. *Journal of Vibration and Shock*, *29*, 237–241.
268. Chen, B., Chen, Z. W., Sun, Y. Z., & Zhao, S. L. (2013). Condition assessment on thermal effects of a suspension bridge based on SHM oriented model and data. *Mathematical Problems in Engineering*, 2013, 1–18.
269. Zhang, X., Du, X., & Brownjohn, J. (2012). Frequency modulated empirical mode decomposition method for the identification of instantaneous modal parameters of aeroelastic systems. *Journal of Wind Engineering and Industrial Aerodynamics*, *101*, 43–52.
270. Yi, J., Zhang, J. W., & Li, Q. S. (2013). Dynamic characteristics and wind-induced responses of a super-tall building during typhoons. *Journal of Wind Engineering and Industrial Aerodynamics*, *121*, 116–130.
271. Wu, Z., & Huang, N. E. (2009). Ensemble empirical mode decomposition: A noise-assisted data analysis method. *Advances in Adaptive Data Analysis*, *1*(01), 1–41.
272. Liu, T. Y., Chiang, W. L., Chen, C. W., Hsu, W. K., Lu, L. C., & Chu, T. J. (2014). Identification and monitoring of bridge health from ambient vibration data. *Journal of Vibration and Control*, *20*(10), 1604–1604. (Retraction of vol. 17, p. 589, 2011).
273. Torres, M. E., Colominas, M. A., Schlotthauer, G., & Flandrin, P. (2011, May). A complete ensemble empirical mode decomposition with adaptive noise. In *2011 IEEE International Conference on Acoustics, Speech and Signal Processing (ICASSP)*, Prague (pp. 4144–4147). IEEE.
274. Wang, T., Zhang, M., Yu, Q., & Zhang, H. (2012). Comparing the applications of EMD and EEMD on time-frequency analysis of seismic signal. *Journal of Applied Geophysics*, *83*, 29–34.
275. Lin, J. W. (2011). A hybrid algorithm based on EEMD and EMD for multi-mode signal processing. *Structural Engineering and Mechanics*, *39*(6), 813–831.
276. Amiri, G. G., & Darvishan, E. (2015). Damage detection of moment frames using ensemble empirical mode decomposition and clustering techniques. *KSCE Journal of Civil Engineering*, *19*(5), 1302–1311.
277. Lei, Y., Liu, Z., Ouazri, J., & Lin, J. (2017). A fault diagnosis method of rolling element bearings based on CEEMDAN. *Proceedings of the Institution of Mechanical Engineers, Part C: Journal of Mechanical Engineering Science*, *231*(10), 1804–1815.
278. Liu, B., Riemenschneider, S., & Xu, Y. (2006). Gearbox fault diagnosis using empirical mode decomposition and Hilbert spectrum. *Mechanical Systems and Signal Processing*, *20*(3), 718–734.

279. Mohanty, S., Gupta, K. K., & Raju, K. S. (2016). Vibro-acoustic fault analysis of bearing using FFT, EMD, EEMD and CEEMDAN and their implications. In P. Soh, W. Woo, H. Sulaiman, M. Othman, & M. Saat (eds.), *Advances in Machine Learning and Signal Processing* (pp. 281–292). Cham: Springer.
280. Lei, Y., He, Z., & Zi, Y. (2009). Application of the EEMD method to rotor fault diagnosis of rotating machinery. *Mechanical Systems and Signal Processing*, *23*(4), 1327–1338.
281. Georgoulas, G., Loutas, T., Stylios, C. D., & Kostopoulos, V. (2013). Bearing fault detection based on hybrid ensemble detector and empirical mode decomposition. *Mechanical Systems and Signal Processing*, *41*(1–2), 510–525.
282. Han, J., & Van der Baan, M. (2013). Empirical mode decomposition for seismic time-frequency analysis. *Geophysics*, *78*(2), O9–O19.
283. Mousavi, A. A., Zhang, C., Masri, S. F., & Gholipour, G. (2021). Structural damage detection method based on the complete ensemble empirical mode decomposition with adaptive noise: A model steel truss bridge case study. *Structural Health Monitoring*. doi: 10.1177/14759217211013535.
284. Mousavi, A. A., Zhang, C., Masri, S. F., & Gholipour, G. (2021). Damage detection and characterization of a scaled model steel truss bridge using combined complete ensemble empirical mode decomposition with adaptive noise and multiple signal classification approach. *Structural Health Monitoring*. doi: 10.1177/14759217211045901.
285. Xiao, F., Chen, G. S., Zatar, W., & Hulsey, J. L. (2021). Signature extraction from the dynamic responses of a bridge subjected to a moving vehicle using complete ensemble empirical mode decomposition. *Journal of Low Frequency Noise, Vibration and Active Control*, *40*(1), 278–294.
286. Li, Y., Chen, X., & Yu, J. (2019). A hybrid energy feature extraction approach for ship-radiated noise based on CEEMDAN combined with energy difference and energy entropy. *Processes*, *7*(2), 69.
287. Lv, Y., Yuan, R., Wang, T., Li, H., & Song, G. (2018). Health degradation monitoring and early fault diagnosis of a rolling bearing based on CEEMDAN and improved MMSE. *Materials*, *11*(6), 1009.
288. Kuai, M., Cheng, G., Pang, Y., & Li, Y. (2018). Research of planetary gear fault diagnosis based on permutation entropy of CEEMDAN and ANFIS. *Sensors*, *18*(3), 782.
289. Sifuzzaman, M., Islam, M. R. & Ali, M. Z. (2009). Application of wavelet transform and its advantages compared to Fourier transform. *Journal of Physical Sciences*, *13*, 121–134.
290. Okafor, A. C., & Dutta, A. (2000). Structural damage detection in beams by wavelet transforms. *Smart Materials and Structures*, *9*(6), 906.
291. Yoon, D. J., Weiss, W. J., & Shah, S. P. (2000). Assessing damage in corroded reinforced concrete using acoustic emission. *Journal of Engineering Mechanics*, *126*(3), 273–283.
292. Melhem, H., & Kim, H. (2003). Damage detection in concrete by Fourier and wavelet analyses. *Journal of Engineering Mechanics*, *129*(5), 571–577.
293. Park, S., Inman, D. J., Lee, J. J., & Yun, C. B. (2008). Piezoelectric sensor-based health monitoring of railroad tracks using a two-step support vector machine classifier. *Journal of Infrastructure Systems*, *14*(1), 80–88.

294. Young Noh, H., Krishnan Nair, K., Lignos, D. G., & Kiremidjian, A. S. (2011). Use of wavelet-based damage-sensitive features for structural damage diagnosis using strong motion data. *Journal of Structural Engineering*, *137*(10), 1215–1228.
295. Su, W. C., Liu, C. Y., & Huang, C. S. (2014). Identification of instantaneous modal parameter of time-varying systems via a wavelet-based approach and its application. *Computer-Aided Civil and Infrastructure Engineering*, *29*(4), 279–298.
296. Li, S., Li, H., Liu, Y., Lan, C., Zhou, W., & Ou, J. (2014). SMC structural health monitoring benchmark problem using monitored data from an actual cable-stayed bridge. *Structural Control and Health Monitoring*, *21*(2), 156–172.
297. Gaviria, C. A., & Montejo, L. A. (2016). Output-only identification of the modal and physical properties of structures using free vibration response. *Earthquake Engineering and Engineering Vibration*, *15*(3), 575–589.
298. Gholizad, A., & Safari, H. (2016). Two-dimensional continuous wavelet transform method for multidamage detection of space structures. *Journal of Performance of Constructed Facilities*, *30*(6), 04016064.
299. Shahsavari, V., Chouinard, L., & Bastien, J. (2017). Wavelet-based analysis of mode shapes for statistical detection and localization of damage in beams using likelihood ratio test. *Engineering Structures*, *132*, 494–507.
300. Abdulkareem, M., Bakhary, N., Vafaei, M., Noor, N. M., & Padil, K. H. (2018). Non-probabilistic wavelet method to consider uncertainties in structural damage detection. *Journal of Sound and Vibration*, *433*, 77–98.
301. Wang, S., Li, J., Luo, H., & Zhu, H. (2019). Damage identification in underground tunnel structures with wavelet based residual force vector. *Engineering Structures*, *178*, 506–520.
302. Pan, H., Azimi, M., Yan, F., & Lin, Z. (2018). Time-frequency-based data-driven structural diagnosis and damage detection for cable-stayed bridges. *Journal of Bridge Engineering*, *23*(6), 04018033.
303. Karami-Mohammadi, R., Mirtaheri, M., Salkhordeh, M., & Hariri-Ardebili, M. A. (2020). Vibration anatomy and damage detection in power transmission towers with limited sensors. *Sensors*, *20*(6), 1731.
304. Hou, Z., Noori, M., & Amand, R. S. (2000). Wavelet-based approach for structural damage detection. *Journal of Engineering Mechanics*, *126*(7), 677–683.
305. Hera, A., & Hou, Z. (2004). Application of wavelet approach for ASCE structural health monitoring benchmark studies. *Journal of Engineering Mechanics*, *130*(1), 96–104.
306. Ovanesova, A. V., & Suarez, L. E. (2004). Applications of wavelet transforms to damage detection in frame structures. *Engineering Structures*, *26*(1), 39–49.
307. Hoseini Vaez, S. R., & Tabaei Aghdaei, S. S. (2019). Effect of the frequency content of earthquake excitation on damage detection in steel frames. *Journal of Rehabilitation in Civil Engineering*, *7*(1), 124–140.
308. Feng, X., Zhang, X., Sun, C., Motamedi, M., & Ansari, F. (2014). Stationary wavelet transform method for distributed detection of damage by fiber-optic sensors. *Journal of Engineering Mechanics*, *140*(4), 04013004.

309. Lucero, J., & Taha, M. R. (2005 Orlando, Florida, USA). A wavelet-aided fuzzy damage detection algorithm for structural health monitoring. In *Proceedings of the 23rd International Modal Analysis Conference (IMAX XXIII)*, Paper (No. 78).
310. Djebala, A., Ouelaa, N., Benchaabane, C., & Laefer, D. F. (2012). Application of the wavelet multi-resolution analysis and Hilbert transform for the prediction of gear tooth defects. *Meccanica*, *47*(7), 1601–1612.
311. Datta, A., Mavroidis, C., Krishnasamy, J., & Hosek, M. (2007, July). Neural netowrk based fault diagnostics of industrial robots using Wavelt multi-resolution analysis. In *2007 American Control Conference*, Boston, MA (pp. 1858–1863). IEEE.
312. Zhang, W., Sun, L., & Zhang, L. (2018). Local damage identification method using finite element model updating based on a new wavelet damage function. *Advances in Structural Engineering*, *21*(10), 1482–1494.
313. Saltari, F., Dessi, D., & Mastroddi, F. (2021). Mechanical systems virtual sensing by proportional observer and multi-resolution analysis. *Mechanical Systems and Signal Processing*, *146*, 107003.
314. Sun, Z., & Chang, C. C. (2002). Structural damage assessment based on wavelet packet transform. *Journal of Structural Engineering*, *128*(10), 1354–1361.
315. Yam, L. H., Yan, Y. J., & Jiang, J. S. (2003). Vibration-based damage detection for composite structures using wavelet transform and neural network identification. *Composite Structures*, *60*(4), 403–412.
316. Han, J. G., Ren, W. X., & Sun, Z. S. (2005). Wavelet packet based damage identification of beam structures. *International Journal of Solids and Structures*, *42*(26), 6610–6627.
317. Ren, W. X., Sun, Z. S., Xia, Y., Hao, H., & Deeks, A. J. (2008). Damage identification of shear connectors with wavelet packet energy: Laboratory test study. *Journal of Structural Engineering*, *134*(5), 832–841.
318. Asgarian, B., Aghaeidoost, V., & Shokrgozar, H. R. (2016). Damage detection of jacket type offshore platforms using rate of signal energy using wavelet packet transform. *Marine Structures*, *45*, 1–21.
319. Chan, C. K., Loh, C. H., & Wu, T. H. (2015, April). Damage detection and quantification in a structural model under seismic excitation using time-frequency analysis. In *Structural Health Monitoring and Inspection of Advanced Materials, Aerospace, and Civil Infrastructure 2015* (Vol. 9437, p. 94372L). International Society for Optics and Photonics. doi: 10.1117/12.2083924.
320. Nie, Z., Guo, E., & Ma, H. (2019). Structural damage detection using wavelet packet transform combining with principal component analysis. *International Journal of Lifecycle Performance Engineering*, *3*(2), 149–170.
321. Daubechies, I., Lu, J., & Wu, H. T. (2011). Synchrosqueezed wavelet transforms: An empirical mode decomposition-like tool. *Applied and Computational Harmonic Analysis*, *30*(2), 243–261.
322. Wen, J., Gao, H., Li, S., Zhang, L., He, X., & Liu, W. (2015, October). Fault diagnosis of ball bearings using synchrosqueezed wavelet transforms and SVM. In *2015 Prognostics and System Health Management Conference (PHM)* (pp. 1–6). Beijing: IEEE.

323. Perez-Ramirez, C. A., Amezquita-Sanchez, J. P., Adeli, H., Valtierra-Rodriguez, M., Camarena-Martinez, D., & Romero-Troncoso, R. J. (2016). New methodology for modal parameters identification of smart civil structures using ambient vibrations and synchrosqueezed wavelet transform. *Engineering Applications of Artificial Intelligence*, *48*, 1–12.
324. Amezquita-Sanchez, J. P., & Adeli, H. (2015). Synchrosqueezed wavelet transform-fractality model for locating, detecting, and quantifying damage in smart highrise building structures. *Smart Materials and Structures*, *24*(6), 065034.
325. Wang, J., Huo, L., Liu, C., & Song, G. (2021). A new acoustic emission damage localization method using synchrosqueezed wavelet transforms picker and time-order method. *Structural Health Monitoring*, *20*(6), 2917–2935.
326. Li, D., Wang, Y., Yan, W. J., & Ren, W. X. (2021). Acoustic emission wave classification for rail crack monitoring based on synchrosqueezed wavelet transform and multi-branch convolutional neural network. *Structural Health Monitoring*, *20*(4), 1563–1582.
327. Su, C., Jiang, M., Liang, J., Tian, A., Sun, L., Zhang, L., ... & Sui, Q. (2020). Damage assessments of composite under the environment with strong noise based on synchrosqueezing wavelet transform and stack autoencoder algorithm. *Measurement*, *156*, 107587.
328. Gilles, J. (2013). Empirical wavelet transform. *IEEE Transactions on Signal Processing*, *61*, 3999–4010.
329. Gilles, J., & Heal, K. (2014). A parameterless scale-space approach to find meaningful modes in histograms-Application to image and spectrum segmentation. *International Journal of Wavelets, Multiresolution and Information Processing*, *12*(06), 1450044.
330. Zheng, J., Pan, H., Yang, S., & Cheng, J. (2017). Adaptive parameterless empirical wavelet transform based time-frequency analysis method and its application to rotor rubbing fault diagnosis. *Signal Processing*, *130*, 305–314.
331. Liu, T., Luo, Z., Huang, J., & Yan, S. (2018). A comparative study of four kinds of adaptive decomposition algorithms and their applications. *Sensors*, *18*(7), 2120.
332. Kedadouche, M., Thomas, M., & Tahan, A. (2016). A comparative study between empirical wavelet transforms and empirical mode decomposition methods: Application to bearing defect diagnosis. *Mechanical Systems and Signal Processing*, *81*, 88–107.
333. Amezquita-Sanchez, J. P., & Adeli, H. (2015). A new music-empirical wavelet transform methodology for time-frequency analysis of noisy nonlinear and non-stationary signals. *Digital Signal Processing*, *45*, 55–68.
334. Reddy, G. R. S., & Rao, R. (2016). Empirical wavelet transform based approach for extraction of fundamental component and estimation of time-varying power quality indices in power quality disturbances. *International Journal of Signal Processing, Image Processing and Pattern Recognition*, *9*, 161–180.
335. Thirumala, K., Jain, T., & Umarikar, A. C. (2017). Visualizing time-varying power quality indices using generalized empirical wavelet transform. *Electric Power Systems Research*, *143*, 99–109.

336. Liu, T., Li, J., Cai, X., & Yan, S. (2018). A time-frequency analysis algorithm for ultrasonic waves generating from a debonding defect by using empirical wavelet transform. *Applied Acoustics*, *131*, 16–27.
337. Yuan, M., Sadhu, A., & Liu, K. (2018). Condition assessment of structure with tuned mass damper using empirical wavelet transform. *Journal of Vibration and Control*, *24*(20), 4850–4867.
338. Xin, Y., Hao, H., & Li, J. (2019). Operational modal identification of structures based on improved empirical wavelet transform. *Structural Control and Health Monitoring*, *26*(3), e2323.
339. Amezquita-Sanchez, J. P., Park, H. S., & Adeli, H. (2017). A novel methodology for modal parameters identification of large smart structures using MUSIC, empirical wavelet transform, and Hilbert transform. *Engineering Structures*, *147*, 148–159.
340. Mousavi, A. A., Zhang, C., Masri, S. F., & Gholipour, G. (2021). Damage detection and localization of a steel truss bridge model subjected to impact and white noise excitations using empirical wavelet transform neural network approach. *Measurement*, *185*, 110060.
341. Yadav, S. K., Banerjee, S., & Kundu, T. (2013). On sequencing the feature extraction techniques for online damage characterization. *Journal of Intelligent Material Systems and Structures*, *24*(4), 473–483.
342. Amjad, U., Yadav, S. K., & Kundu, T. (2015). Detection and quantification of diameter reduction due to corrosion in reinforcing steel bars. *Structural Health Monitoring*, *14*(5), 532–543.
343. Pakrashi, V., & Ghosh, B. (2009). Application of S transform in structural health monitoring. In *7th International Symposium on Nondestructive Testing in Civil Engineering (NDTCE)*. Nantes, France, 30 Jun–3 Jul 2009.
344. Ditommaso, R., Mucciarelli, M., Parolai, S., & Picozzi, M. (2012). Monitoring the structural dynamic response of a masonry tower: Comparing classical and time-frequency analyses. *Bulletin of Earthquake Engineering*, *10*(4), 1221–1235.
345. Amini Tehrani, H., Bakhshi, A., & Akhavat, M. (2017). An effective approach to structural damage localization in flexural members based on generalized S-transform. *Scientia Iranica*, *26*(6), 3125–3139.
346. Liu, N., Xi, J., Zhang, X., & Liu, Z. (2017, August). Damage detection of simply supported reinforced concrete beam by S transform. In *IOP Conference Series: Earth and Environmental Science*, Zhuhai (Vol. 81, No. 1, p. 012133). IOP Publishing.
347. Ponzo, F. C., Ditommaso, R., Auletta, G., Iacovino, C., Mossucca, A., Nigro, A., & Nigro, D. (2014). Localization of damage occurred on framed structures: Analysis of the geometric characteristics of the fundamental mode shape. *Earthquake Engineering*, *31*(3).
348. Ponzo, F. C., Ditommaso, R., Auletta, G., & De Muro, A. (2013 Cape Town, South Africa, August). Damage detection on Reinforced Concrete Framed Structures using a band-variable filter. In *Research and Applications in Structural Engineering, Mechanics and Computation-Proceedings of the 5th International Conference on Structural Engineering, Mechanics and Computation, SEMC 2013* (pp. 2303–2307).

349. Iacovino, C., Ditommaso, R., Limongelli, M. P., & Carlo, F. (2016). Comparison of the performance of two different approaches for damage detection on framed structures. In *8th European Workshop on Structural Health Monitoring, EWSHM 2016*, Spain (Vol. 3, pp. 2017–2026). NDT. net.
350. Ditommaso, R., & Ponzo, F. C. (2015). Automatic evaluation of the fundamental frequency variations and related damping factor of reinforced concrete framed structures using the Short Time Impulse Response Function (STIRF). *Engineering Structures*, *82*, 104–112.
351. Brown, R. A., & Frayne, R. (2008, August). A fast discrete S-transform for biomedical signal processing. In *2008 30th Annual International Conference of the IEEE Engineering in Medicine and Biology Society*, Vancouver, BC (pp. 2586–2589). IEEE.
352. Ghahremani, B., Bitaraf, M., Ghorbani-Tanha, A. K., & Fallahi, R. (2021, February). Structural damage identification based on fast S-transform and convolutional neural networks. In *Structures* (Vol. 29, pp. 1199–1209). Elsevier. 10.1016/j.istruc.2020.11.068.
353. Cohen, L. (1966). Generalized phase-space distribution functions. *Journal of Mathematical Physics*, *7*(5), 781–786.
354. Staszewski, W. J., Worden, K., & Tomlinson, G. R. (1997). Time-frequency analysis in gearbox fault detection using the Wigner-Ville distribution and pattern recognition. *Mechanical Systems and Signal Processing*, *11*(5), 673–692.
355. Claasen, T. A. C. M., & Mecklenbräuker, W. (1980). Time-frequency signal analysis. *Philips Journal of Research*, *35*(6), 372–389.
356. Baydar, N., & Ball, A. (2001). A comparative study of acoustic and vibration signals in detection of gear failures using Wigner-Ville distribution. *Mechanical Systems and Signal Processing*, *15*(6), 1091–1107.
357. Zou, J., & Chen, J. (2004). A comparative study on time-frequency feature of cracked rotor by Wigner-Ville distribution and wavelet transform. *Journal of Sound and Vibration*, *276*(1–2), 1–11.
358. Gillich, G. R., & Praisach, Z. I. (2014). Modal identification and damage detection in beam-like structures using the power spectrum and time-frequency analysis. *Signal Processing*, *96*, 29–44.
359. Dai, D., & He, Q. (2014). Structure damage localization with ultrasonic guided waves based on a time-frequency method. *Signal Processing*, *96*, 21–28.
360. Katunin, A. (2020). Damage identification and quantification in beams using Wigner-Ville distribution. *Sensors*, *20*(22), 6638.
361. Qatu, K. M., Abdelgawad, A., & Yelamarthi, K. (2016, March). Structure damage localization using a reliable wave damage detection technique. In *2016 International Conference on Electrical, Electronics, and Optimization Techniques (ICEEOT)*, Chennai (pp. 1959–1962). IEEE.
362. Wu, J., Ma, Z., & Zhang, Y. (2017). A time-frequency research for ultrasonic guided wave generated from the debonding based on a novel time-frequency analysis technique. *Shock and Vibration*, 2017, 1–11.
363. Lin, Y. Z., Nie, Z. H., Ma, H. W. (2017). Structural damage detection with automatic feature-extraction through deep learning. *Computer-Aided Civil and Infrastructure Engineering*, *32* (12), 1025–1046.

364. Wang, Z., Cha, Y. J. (2018). Automated damage-sensitive feature extraction using unsupervised convolutional neural networks. In *Sensors and Smart Structures Technologies for Civil, Mechanical, and Aerospace Systems 2018* (vol. 10598, p. 105981J). International Society for Optics and Photonics. doi: 10.1117/12.2295966.
365. Wu, R. T., Jahanshahi, M. R. (2019). Deep convolutional neural network for structural dynamic response estimation and system identification. *Journal of Engineering Mechanics*, *145*(1), 04018125.
366. Pathirage, J. Li, L. Li, Hao, H., Liu, W., Ni, P. (2018). Structural damage identification based on autoencoder neural networks and deep learning. *Engineering Structures*, *172*, 13–28.
367. Lopez-Pacheco, M., Morales-Valdez, J., Yu, W. (2020). Frequency domain CNN and dissipated energy approach for damage detection in building structures. *Soft Computing*, *24*(20), 15821–15840.
368. Pan, H., Azimi, M., Gui, G., Yan, F., Lin, Z. (2017). Vibration-based support vector machine for structural health monitoring. In *International Conference on Experimental Vibration Analysis for Civil Engineering Structures* (pp. 167–178). Cham: Springer. 10.1007/978-3-319-67443-8_14.
369. Pan, X., Yang, T. Y. (2020). Postdisaster image-based damage detection and repair cost estimation of reinforced concrete buildings using dual convolutional neural networks. *Computer-Aided Civil and Infrastructure Engineering*, *35*(5), 495–510.
370. Bao, Y., Tang, Z., Li, H., Zhang, Y. (2019). Computer vision and deep learning-based data anomaly detection method for structural health monitoring, *Structural Health Monitoring*, *18*(2), 401–421.
371. Tang, Z., Chen, Z., Bao, Y., Li, H. (2019). Convolutional neural network-based data anomaly detection method using multiple information for structural health monitoring. *Structural Control and Health Monitoring*, *26*(1), e2296.
372. Khodabandehlou, H., Pekcan, G., & Fadali, M. S. (2019). Vibration-based structural condition assessment using convolution neural networks. *Structural Control and Health Monitoring*, *26*(2), e2308.
373. Li, S., & Sun, L. (2020). Detectability of bridge-structural damage based on fiber-optic sensing through deep-convolutional neural networks. *Journal of Bridge Engineering*, *25*(4), 04020012.
374. Li, S., Zuo, X., Li, Z., & Wang, H. (2020). Applying deep learning to continuous bridge deflection detected by fiber optic gyroscope for damage detection. *Sensors*, *20*(3), 911.
375. Hung, D. V., Hung, H. M., Anh, P. H., & Thang, N. T. (2020). Structural damage detection using hybrid deep learning algorithm. *Journal of Science and Technology in Civil Engineering (STCE)-NUCE*, *14*(2), 53–64.
376. Ding, Z., Li, J., & Hao, H. (2020). Structural damage identification by sparse deep belief network using uncertain and limited data. *Structural Control and Health Monitoring*, *27*(5), e2522.
377. Fan, G., Li, J., & Hao, H. (2020). Vibration signal denoising for structural health monitoring by residual convolutional neural networks. *Measurement*, *157*, 107651.

378. Ying, Y., Garrett Jr., J. H., Oppenheim, I. J., Soibelman, L., Harley, J. B., Shi, J., & Jin, Y. (2013). Toward data-driven structural health monitoring: Application of machine learning and signal processing to damage detection. *Journal of Computing in Civil Engineering*, 27(6), 667–680.
379. Shang, Z., Sun, L., Xia, Y., & Zhang, W. (2021). Vibration-based damage detection for bridges by deep convolutional denoising autoencoder. *Structural Health Monitoring*, *20*(4), 1880–1903.
380. Rafiei, M. H., & Adeli, H. (2017). A novel machine learning-based algorithm to detect damage in high-rise building structures. *Structural Design of Tall and Special Buildings*, *26*(18), e1400.
381. Yan, Z., Miyamoto, A., & Jiang, Z. (2009). Frequency slice wavelet transform for transient vibration response analysis. *Mechanical Systems and Signal Processing*, 23(5), 1474–1489.
382. Ruocci, G., Cumunel, G., Le, T., Argoul, P., Point, N., & Dieng, L. (2014). Damage assessment of pre-stressed structures: A SVD-based approach to deal with time-varying loading. *Mechanical Systems and Signal Processing*, *47*(1–2), 50–65.
383. Cheraghi, N., Zou, G. P., & Taheri, F. (2005). Piezoelectric-based degradation assessment of a pipe using fourier and wavelet analyses. *Computer-Aided Civil and Infrastructure Engineering*, 20(5), 369–382.
384. Ruocci, G., Quattrone, A., & De Stefano, A. (2011, July). Multi-domain feature selection aimed at the damage detection of historical bridges. In *Journal of Physics: Conference Series* (Vol. 305, No. 1, p. 012106). IOP Publishing.
385. Maia, N. M. M., Silva, J. M. M., & Sampaio, R. P. C. (1997). Localization of damage using curvature of the frequency-response-functions. In *Proceedings of the 15th International Modal Analysis Conference*, Orlando, FL (Vol. 3089, p. 942).
386. Esfandiari, A., Nabiyan, M. S., & Rofooei, F. R. (2020). Structural damage detection using principal component analysis of frequency response function data. *Structural Control and Health Monitoring*, *27*(7), e2550.
387. Kao, C. Y., Chen, X. Z., & Hung, S. L. (2020). A displacement frequency response function-based approach for locating damage to building structures. *Advances in Civil Engineering*, 2020, 1–23.
388. Dilena, M., Limongelli, M. P., & Morassi, A. (2015). Damage localization in bridges via the FRF interpolation method. *Mechanical Systems and Signal Processing*, *52*, 162–180.
389. Kim, M. J., & Eun, H. C. (2017). Identification of damage-expected members of truss structures using frequency response function. *Advances in Mechanical Engineering*, *9*(1). doi: 10.1177/1687814016687911.
390. Huang, N. E., Shen, Z., Long, S. R., Wu, M. C., Shih, H. H., Zheng, Q., ... & Liu, H. H. (1998). The empirical mode decomposition and the Hilbert spectrum for nonlinear and non-stationary time series analysis. *Proceedings of the Royal Society of London. Series A: Mathematical, Physical and Engineering Sciences*, *454*(1971), 903–995.
391. Quek, S. T., Tua, P. S., & Wang, Q. (2003). Detecting anomalies in beams and plate based on the Hilbert-Huang transform of real signals. *Smart Materials and Structures*, *12*(3), 447.

392. Yang, J. N., Lei, Y., Lin, S., & Huang, N. (2004). Hilbert-Huang based approach for structural damage detection. *Journal of Engineering Mechanics*, *130*(1), 85–95.
393. Liu, J., Wang, X., Yuan, S., & Li, G. (2006). On Hilbert-Huang transform approach for structural health monitoring. *Journal of Intelligent Material Systems and Structures*, *17*(8–9), 721–728.
394. Yan, B., & Miyamoto, A. (2006). A comparative study of modal parameter identification based on wavelet and Hilbert-Huang transforms. *Computer-Aided Civil and Infrastructure Engineering*, *21*(1), 9–23.
395. Li, H., Deng, X., & Dai, H. (2007). Structural damage detection using the combination method of EMD and wavelet analysis. *Mechanical Systems and Signal Processing*, *21*(1), 298–306.
396. Cheraghi, N., & Taheri, F. (2007). A damage index for structural health monitoring based on the empirical mode decomposition. *Journal of Mechanics of Materials and Structures*, *2*(1), 43–61.
397. Rezaei, D., & Taheri, F. (2009). Experimental validation of a novel structural damage detection method based on empirical mode decomposition. *Smart Materials and Structures*, *18*(4), 045004.
398. Hsu, W. K., Chiou, D. J., Chen, C. W., Liu, M. Y., Chiang, W. L., & Huang, P. C. (2014). Sensitivity of initial damage detection for steel structures using the Hilbert-Huang transform method (Retraction of vol. 19, p. 857, 2013). *Journal of Vibration and Control*, *20*(10), 1603–1603.
399. Carbajo, E. S., Carbajo, R. S., Mc Goldrick, C., & Basu, B. (2014). ASDAH: An automated structural change detection algorithm based on the Hilbert-Huang transform. *Mechanical Systems and Signal Processing*, *47*(1–2), 78–93.
400. Dushyanth, N. D., Suma, M. N., & Latte, M. V. (2016). Detection and localization of damage using empirical mode decomposition and multilevel support vector machine. *Applied Physics A*, *122*, 1–9.
401. Salvino, L. W., Pines, D. J., Todd, M., & Nichols, J. M. (2005). EMD and instantaneous phase detection of structural damage. In *Hilbert-Huang Transform and Its Applications* (pp. 227–262). doi: 10.1142/9789812703347_0011.
402. Pakrashi, V., Basu, B., & O'Connor, A. (2007). Structural damage detection and calibration using a wavelet-kurtosis technique. *Engineering Structures*, *29*(9), 2097–2108.
403. Xu, Y., & Zhang, H. (2009). Recent mathematical developments on empirical mode decomposition. *Advances in Adaptive Data Analysis*, *1*(04), 681–702.
404. Flandrin, P., Rilling, G., & Goncalves, P. (2004). Empirical mode decomposition as a filter bank. *IEEE Signal Processing Letters*, *11*(2), 112–114.
405. Cheraghi, N., Zou, G. P., & Taheri, F. (2005). Piezoelectric-based degradation assessment of a pipe using fourier and wavelet analyses. *Computer-Aided Civil and Infrastructure Engineering*, *20*(5), 369–382.
406. Huang, N. E., Shen, Z., & Long, S. R., et al. (1998). The empirical mode decomposition and the Hilbert spectrum for nonlinear and non-stationary time series analysis. In: *Proceedings of the Royal Society of London. Series A: Mathematical, Physical and Engineering Sciences* (vol. 454, pp. 903–995), London, 8 March 1998.
407. Huang, N. E., Long, S. R., & Shen, Z. (1996). The mechanism for frequency downshift in nonlinear wave evolution. *Advances in Applied Mechanics*, *32*, 59–117C.

408. Mohammadi Ghazi, R., & Büyüköztürk, O. (2016). Damage detection with small data set using energy-based nonlinear features. *Structural Control and Health Monitoring*, *23*(2), 333–348.
409. Zhu, X., Cao, M., Ostachowicz, W., & Xu, W. (2019). Damage identification in bridges by processing dynamic responses to moving loads: Features and evaluation. *Sensors*, *19*(3), 463.
410. Lee, S. G., Yun, G. J., & Shang, S. (2014). Reference-free damage detection for truss bridge structures by continuous relative wavelet entropy method. *Structural Health Monitoring*, *13*(3), 307–320.
411. Delgadillo, R. M., & Casas, J. R. (2020). Non-modal vibration-based methods for bridge damage identification. *Structure and Infrastructure Engineering*, *16*(4), 676–697.
412. Pai, P. F., Huang, L., Hu, J., & Langewisch, D. R. (2008). Time-frequency method for nonlinear system identification and damage detection. *Structural Health Monitoring*, *7*(2), 103–127.
413. Huang, N. E., Huang, K., & Chiang, W. L. (2005). Chapter 12: HHT-based bridge structural health-monitoring method: Hilbert-Huang transform and its applications. In: N. E. Huang & S. S. P. Shen (Eds.), *Hilbert-Huang Transform and its Applications* (pp. 263–287). Hackensack, NJ: World Scientific.
414. Huang, F. L., Wang, X. M., Chen, Z. Q., & He, X. H. (2006). Application of Hilbert Huang transform analysis to health monitoring of Nanjing Yangtze River Bridge. In *Proceedings of the Second International Conference on Structural Health Monitoring of Intelligent Infrastructure*. London: Taylor & Francis (pp. 731–736).
415. Chen J. (2002). Hilbert-Huang transform for damping ratio identification of structures with closely spaced modes of vibration. In *Proceedings of the International Conference on Advances in Building Technology*, Hong Kong, China, 4–6 December 2002. New York: Springer (pp. 1107–1114).
416. Chen, J., & Xu, Y. L. (2002). Identification of modal damping ratios of structures with closely spaced modal frequencies. *Structural Engineering and Mechanics, An Int'l Journal*, *14*(6), 417–434.
417. Huang, N. E., & Attoh-Okine, N. O. (2005). *The Hilbert-Huang Transform in Engineering*. Boca Raton, FL: CRC Press.
418. Yang, J. N., Lei, Y., Pan, S., & Huang, N. (2003). System identification of linear structures based on Hilbert-Huang spectral analysis. Part 1: Normal modes. *Earthquake Engineering & Structural Dynamics*, *32*(9), 1443–1467.
419. Yang, J. N., Lei, Y., Pan, S., & Huang, N. (2003). System identification of linear structures based on Hilbert-Huang spectral analysis. Part 2: Complex modes. *Earthquake Engineering & Structural Dynamics*, *32*(10), 1533–1554.
420. Gholipour, G., Zhang, C., & Mousavi, A. A. (2020). Nonlinear numerical analysis and progressive damage assessment of a cable-stayed bridge pier subjected to ship collision. *Marine Structures*, *69*, 102662.
421. Lei, Y., Liu, Z., Ouazri, J., & Lin, J. (2017). A fault diagnosis method of rolling element bearings based on CEEMDAN. *Proceedings of the Institution of Mechanical Engineers, Part C: Journal of Mechanical Engineering Science*, *231*(10), 1804–1815.
422. Lei, Y., He, Z., & Zi, Y. (2009). Application of the EEMD method to rotor fault diagnosis of rotating machinery. *Mechanical Systems and Signal Processing*, *23*(4), 1327–1338.

423. Georgoulas, G., Loutas, T., Stylios, C. D., & Kostopoulos, V. (2013). Bearing fault detection based on hybrid ensemble detector and empirical mode decomposition. *Mechanical Systems and Signal Processing*, *41*(1–2), 510–525.
424. Han, J., & van der Baan, M. (2013). Empirical mode decomposition for seismic time-frequency analysis. *Geophysics*, *78*(2), O9–O19.
425. Gilles, J. (2013). Empirical wavelet transform. *IEEE Transactions on Signal Processing*, *61*(16), 3999–4010.
426. Stoica, P., & Moses, R. L. (2005). *Spectral Analysis of Signals* (Vol. 452, pp. 25–26). Upper Saddle River, NJ: Pearson Prentice Hall.
427. Gao, Y., & Spencer Jr, B. F. (2007). Experimental verification of a distributed computing strategy for structural health monitoring. *Smart Structures and Systems, An International Journal*, *3*(4), 455–474.
428. Tributsch, A., & Adam, C. (2018). An enhanced energy vibration-based approach for damage detection and localization. *Structural Control and Health Monitoring*, *25*(1), e2047.
429. Cheraghi, N., & Taheri, F. (2008). Application of the empirical mode decomposition for system identification and structural health monitoring. *International Journal of Applied Mathematics and Engineering Sciences*, 2(1), 61–72.
430. Cheraghi, N., Riley, M. J., & Taheri, F. (2005, October). A novel approach for detection of damage in adhesively bonded joints in plastic pipes based on vibration method using piezoelectric sensors. In *2005 IEEE International Conference on Systems, Man and Cybernetics*, Waikoloa, HI (Vol. 4, pp. 3472–3478). IEEE.
431. Flandrin, P., Rilling, G., & Goncalves, P. (2004). Empirical mode decomposition as a filter bank. *IEEE Signal Processing Letters*, *11*(2), 112–114.
432. Johnson, E. A., Lam, H. F., Katafygiotis, L. S., & Beck, J. L. (2001). A benchmark problem for structural health monitoring and damage detection. In *Structural Control for Civil and Infrastructure Engineering*, London (pp. 317–324).
433. Katafygiotis, L. S., Lam, H. F., & Mickleborough, N. (2000, May). Application of a statistical approach on a benchmark damage detection problem. In *Proceedings of the 14th ASCE Engineering Mechanics Conference*, Austin, TX (pp. 21–24).
434. Dyke, S. J., Caicedo, J. M., & Johnson, E. A. (2000, May). Monitoring of a benchmark structure for damage identification. In *Proceedings of the Engineering Mechanics Speciality Conference*, Austin, TX (pp. 21–24).
435. Bernal, D., & Gunes, B. (2000, May). Observer/Kalman and subspace identification of the UBC benchmark structural model. In *Proceedings of the 14th ASCE Engineering Mechanics Conference*, Austin, TX (pp. 21–24).
436. Lin, S., Yang, J. N., & Zhou, L. (2005). Damage identification of a benchmark building for structural health monitoring. *Smart Materials and Structures*, *14*(3), S162.
437. Reda Taha, M. M. (2010). A neural-wavelet technique for damage identification in the ASCE benchmark structure using phase II experimental data. *Advances in Civil Engineering*, *2010*, 1–13.
438. Figueiredo, E., Park, G., Farrar, C. R., Worden, K., & Figueiras, J. (2011). Machine learning algorithms for damage detection under operational and environmental variability. *Structural Health Monitoring*, *10*(6), 559–572.

439. Civera, M., & Surace, C. (2021). A comparative analysis of signal decomposition techniques for structural health monitoring on an experimental benchmark. *Sensors*, *21*(5), 1825.
440. Figueiredo, E., Park, G., Figueiras, J., Farrar, C., & Worden, K. (2009). Structural health monitoring algorithm comparisons using standard data sets (No. LA-14393). Los Alamos National Lab (LANL), Los Alamos, NM (United States).
441. Jiang, X., & Adeli, H. (2007). Pseudospectra, MUSIC, and dynamic wavelet neural network for damage detection of Highrise buildings. *International Journal for Numerical Methods in Engineering*, *71*(5), 606–629.
442. Kunwar, A., Jha, R., Whelan, M., & Janoyan, K. (2013). Damage detection in an experimental bridge model using Hilbert-Huang transform of transient vibrations. *Structural Control and Health Monitoring*, *20*(1), 1–15.
443. Esmaeel, R. A., Briand, J., & Taheri, F. (2012). Computational simulation and experimental verification of a new vibration-based structural health monitoring approach using piezoelectric sensors. *Structural Health Monitoring*, *11*(2), 237–250.
444. Han, D., Wei, S., Shi, P., Zhang, Y., Gao, K., & Tian, N. (2017). Damage identification of a Derrick steel structure based on the HHT marginal spectrum amplitude curvature difference. *Shock and Vibration*, *2017*, 9.
445. Tua, P. S., Quek, S. T., & Wang, Q. (2004). Detection of cracks in plates using piezo-actuated Lamb waves. *Smart Materials and Structures*, *13*(4), 643.
446. Pines, D. J., & Salvino, L. W. (2009). Damage Detection Using the Hilbert-Huang Transform. Encyclopedia of Structural Health Monitoring.
447. Jha, R., Cross, K., Janoyan, K., Sazonov, E., Fuchs, M., & Krishnamurthy, V. (2006, April). Experimental evaluation of instantaneous phase based index for structural health monitoring. In *Smart Structures and Materials 2006: Smart Structures and Integrated Systems*, San Diego, CA (Vol. 6173, pp. 418–425). SPIE.
448. Pines, D. J., & Salvino, L. W. (2002, July). Health monitoring of one-dimensional structures using empirical mode decomposition and the Hilbert-Huang transform. In *Smart Structures and Materials 2002: Smart Structures and Integrated Systems*, San Diego, CA (Vol. 4701, pp. 127–143). SPIE.
449. Babu, T. R., Srikanth, S., & Sekhar, A. S. (2008). Hilbert-Huang transform for detection and monitoring of crack in a transient rotor. *Mechanical Systems and Signal Processing*, *22*(4), 905–914.
450. Yan, R., & Gao, R. X. (2006). Hilbert-Huang transform-based vibration signal analysis for machine health monitoring. *IEEE Transactions on Instrumentation and Measurement*, *55*(6), 2320–2329.
451. Cheraghi, N., & Taheri, F. (2007). A damage index for structural health monitoring based on the empirical mode decomposition. *Journal of Mechanics of Materials and Structures*, *2*(1), 43–61.
452. Garcia-Perez, A., Romero-Troncoso, R. J., Cabal-Yepez, E., Osornio-Rios, R. A., & Lucio-Martinez, J. A. (2012). Application of high-resolution spectral analysis for identifying faults in induction motors by means of sound. *Journal of Vibration and Control*, *18*(11), 1585–1594.
453. Fan, S., Zhang, A., Sun, H., & Yun, F. (2021). A local TR-MUSIC algorithm for damage imaging of aircraft structures. *Sensors*, *21*(10), 3334.

454. Zhang, C., Li, H., & Hu, H. T. (2021). Aircraft bearing fault diagnosis based on automatic feature engineering. *CIESC Journal*, *72*, 430–436.
455. Kuai, M., Cheng, G., Pang, Y., & Li, Y. (2018). Research of planetary gear fault diagnosis based on permutation entropy of CEEMDAN and ANFIS. *Sensors*, *18*(3), 782.
456. Cheraghi, N., Zou, G. P., & Taheri, F. (2005). Piezoelectric-based degradation assessment of a pipe using Fourier and wavelet analyses. *Computer-Aided Civil and Infrastructure Engineering*, *20*(5), 369–382.
457. Huang, N. E., Shen, Z., Long, S. R., Wu, M. C., Shih, H. H., Zheng, Q., ... & Liu, H. H. (1998). The empirical mode decomposition and the Hilbert spectrum for nonlinear and non-stationary time series analysis. *Proceedings of the Royal Society of London. Series A: Mathematical, Physical and Engineering Sciences*, *454*(1971), 903–995.
458. Huang, N. E., Long, S. R., & Shen, Z. (1996). The mechanism for frequency downshift in nonlinear wave evolution. *Advances in Applied Mechanics*, *32*, 59–117C.
459. Kunwar, A., Jha, R., Whelan, M., & Janoyan, K. (2013). Damage detection in an experimental bridge model using Hilbert-Huang transform of transient vibrations. *Structural Control and Health Monitoring*, *20*(1), 1–15.
460. Huang, N. E., Huang, K., & Chiang, W. L. (2014). HHT-based bridge structural health-monitoring method. In *Hilbert-Huang Transform and Its Applications* (pp. 337–361). doi: 10.1142/9789812703347_0012.
461. Jiang, X., & Adeli, H. (2007). Pseudospectra, MUSIC, and dynamic wavelet neural network for damage detection of Highrise buildings. *International Journal for Numerical Methods in Engineering*, *71*(5), 606–629.
462. Garcia-Perez, A., Romero-Troncoso, R. J., Cabal-Yepez, E., Osornio-Rios, R. A., & Lucio-Martinez, J. A. (2012). Application of high-resolution spectral analysis for identifying faults in induction motors by means of sound. *Journal of Vibration and Control*, *18*(11), 1585–1594.
463. Camarena-Martinez, D., Osornio-Rios, R., Romero-Troncoso, R. J., & Garcia-Perez, A. (2015). Fused empirical mode decomposition and MUSIC algorithms for detecting multiple combined faults in induction motors. *Journal of Applied Research and Technology*, *13*(1), 160–167.
464. Balafas, K., Kiremidjian, A. S., & Rajagopal, R. (2018). The wavelet transform as a Gaussian process for damage detection. *Structural Control and Health Monitoring*, *25*(2), e2087.
465. Ondra, V., Sever, I. A., & Schwingshackl, C. W. (2017). A method for detection and characterization of structural non-linearities using the Hilbert transform and neural networks. *Mechanical Systems and Signal Processing*, *83*, 210–227.
466. Huang, N. E., Shen, Z., Long, S. R., Wu, M. C., Shih, H. H., Zheng, Q., ... & Liu, H. H. (1998). The empirical mode decomposition and the Hilbert spectrum for nonlinear and non-stationary time series analysis. *Proceedings of the Royal Society of London. Series A: Mathematical, Physical and Engineering Sciences*, *454*(1971), 903–995.
467. Gupta, V., & Pachori, R. B. (2021). FBDM based time-frequency representation for sleep stages classification using EEG signals. *Biomedical Signal Processing and Control*, *64*, 102265.

468. Schroeder, J. (1993). Signal processing via Fourier-Bessel series expansion. *Digital Signal Processing*, *3*(2), 112–124.
469. Bhattacharyya, A., Singh, L., & Pachori, R. B. (2018). Fourier-Bessel series expansion based empirical wavelet transform for analysis of non-stationary signals. *Digital Signal Processing*, *78*, 185–196.
470. Bayram, I. (2012). An analytic wavelet transform with a flexible time-frequency covering. *IEEE Transactions on Signal Processing*, *61*(5), 1131–1142.
471. Gupta, V., Bhattacharyya, A., & Pachori, R. B. (2020). Automated identification of epileptic seizures from EEG signals using FBSE-EWT method. In *Biomedical Signal Processing: Advances in Theory, Algorithms and Applications* (pp. 157–179). 10.1007/978-981-13-9097-5_8.
472. Tripathy, R. K., Bhattacharyya, A., & Pachori, R. B. (2019). A novel approach for detection of myocardial infarction from ECG signals of multiple electrodes. *IEEE Sensors Journal*, *19*(12), 4509–4517.
473. Gupta, V., & Pachori, RB (2019). Epileptic seizure identification using entropy of FBSE based EEG rhythms. *Biomedical Signal Processing and Control*, *53*, 101569.
474. Katiyar, R., Gupta, V., & Pachori, R. B. (2019). FBSE-EWT-based approach for the determination of respiratory rate from PPG signals. *IEEE Sensors Letters*, *3*(7), 1–4.
475. Tripathy, RK, Bhattacharyya, A., & Pachori, RB (2019). Localization of myocardial infarction from multi-lead ECG signals using multiscale analysis and convolutional neural network. *IEEE Sensors Journal*, *19* (23), 11437–11448.
476. Gilles, J. (2013). Empirical wavelet transform. *IEEE Transactions on Signal Processing*, *61*(16), 3999–4010.
477. Yan, YJ, Cheng, L., Wu, ZY, & Yam, LH (2007). Development in vibration-based structural damage detection technique. *Mechanical Systems and Signal Processing*, *21*(5), 2198–2211.
478. Gholipour, G., Zhang, C., & Mousavi, A. A. (2018). Effects of axial load on nonlinear response of RC columns subjected to lateral impact load: Ship-pier collision. *Engineering Failure Analysis*, *91*, 397–418.
479. Gholipour, G., Zhang, C., & Mousavi, A. A. (2020). Nonlinear numerical analysis and progressive damage assessment of a cable-stayed bridge pier subjected to ship collision. *Marine Structures*, *69*, 102662.
480. Gholipour, G., Zhang, C., & Mousavi, A. A. (2021). Nonlinear failure analysis of bridge pier subjected to vessel impact combined with blast loads. *Ocean Engineering*, *234*, 109209.
481. Adeli, H. (2001). Neural networks in civil engineering: 1989-2000. *Computer-Aided Civil and Infrastructure Engineering*, *16*(2), 126–142.
482. Nazarko, P., & Ziemiański, L. (2017). Application of artificial neural networks in the damage identification of structural elements. *Computer Assisted Methods in Engineering and Science*, *18*(3), 175–189.
483. Gordan, M., Razak, H. A., Ismail, Z., & Ghaedi, K. (2017). Recent developments in damage identification of structures using data mining. *Latin American Journal of Solids and Structures*, *14*, 2373–2401.
484. Qian, Y., & Mita, A. (2008). Acceleration-based damage indicators for building structures using neural network emulators. *Structural Control and Health Monitoring: The Official Journal of the International Association for Structural Control and Monitoring and of the European Association for the Control of Structures*, *15*(6), 901–920.

485. Masri, S. F., Nakamura, M., Chassiakos, A. G., & Caughey, T. K. (1996). Neural network approach to detection of changes in structural parameters. *Journal of Engineering Mechanics*, *122*(4), 350–360.
486. Masri, S. F., Smyth, A. W., Chassiakos, A. G., Caughey, T. K., & Hunter, N. F. (2000). Application of neural networks for detection of changes in nonlinear systems. *Journal of Engineering Mechanics*, *126*(7), 666–676.
487. Dackermann, U., Smith, WA, & Randall, RB (2014). Damage identification based on response-only measurements using cepstrum analysis and artificial neural networks. *Structural Health Monitoring*, *13*(4), 430–444.
488. Arangio, S., & Bontempi, F. (2015). Structural health monitoring of a cable-stayed bridge with Bayesian neural networks. *Structure and Infrastructure Engineering*, *11*(4), 575–587.
489. Zang, C., Friswell, M. I., & Imregun, M. (2004). Structural damage detection using independent component analysis. *Structural Health Monitoring*, *3*(1), 69–83.
490. Garcia-Perez, A., Amezquita-Sanchez, J. P., Dominguez-Gonzalez, A., Sedaghati, R., Osornio-Rios, R., & Romero-Troncoso, R. J. (2013). Fused empirical mode decomposition and wavelets for locating combined damage in a truss-type structure through vibration analysis. *Journal of Zhejiang University SCIENCE A*, *14*, 615–630.
491. Xun, J., & Yan, S. (2008). A revised Hilbert-Huang transformation based on the neural networks and its application in vibration signal analysis of a deployable structure. *Mechanical Systems and Signal Processing*, *22*(7), 1705–1723.
492. Timmermann, A., & Granger, CW (2004). Efficient market hypothesis and forecasting. *International Journal of Forecasting*, *20*(1), 15–27.
493. Flandrin, P., Rilling, G., & Goncalves, P. (2004). Empirical mode decomposition as a filter bank. *IEEE Signal Processing Letters*, *11*(2), 112–114.
494. Liu, H., Zhou, J., Zheng, Y., Jiang, W., & Zhang, Y. (2018). Fault diagnosis of rolling bearings with recurrent neural network-based autoencoders. *ISA Transactions*, *77*, 167–178.
495. Kedadouche, M., Thomas, M., & Tahan, A. J. M. S. (2016). A comparative study between empirical wavelet transforms and empirical mode decomposition methods: Application to bearing defect diagnosis. *Mechanical Systems and Signal Processing*, *81*, 88–107.
496. Xia, YX, & Zhou, YL (2019). Mono-component feature extraction for condition assessment in civil structures using empirical wavelet transform. *Sensors*, *19* (19), 4280.
497. Huang, N. E., Huang, K., & Chiang, W. L. (2014). HHT-based bridge structural health-monitoring method. In *Hilbert-Huang Transform and Its Applications* (pp. 337–361). doi: 10.1142/9789812703347_0012.
498. Kunwar, A., Jha, R., Whelan, M., & Janoyan, K. (2013). Damage detection in an experimental bridge model using Hilbert-Huang transform of transient vibrations. *Structural Control and Health Monitoring*, *20*(1), 1–15.

Index